Undergraduate Texts in Mathematics

Gordon Whyburn
Edwin Duda

Dynamic Topology

With 20 Figures

Springer-Verlag
New York Heidelberg Berlin

191825

Gordon Whyburn
formerly of
Department of Mathematics
University of Virginia
Charlottesville, Virginia 22903
USA

Edwin Duda
Department of Mathematics
University of Miami
Coral Gables, Florida 33124
USA

Editorial Board
F.W. Gehring
Department of Mathematics
University of Michigan
Ann Arbor, Michigan 48104
USA

P.R. Halmos
Department of Mathematics
Indiana University
Bloomington, Indiana 47401
USA

AMS Subject Classification: 54-01

Library of Congress Cataloging in Publication Data

Whyburn, Gordon Thomas,
  Dynamic topology.

  (Undergraduate texts in mathematics)
  Bibliography: p.
  Includes index.
  1. Topology.   I. Duda, Edwin, joint author.
II. Title.
QA611.W494   514   78-24569

The Appendix: Dynamic Topology by G.T. Whyburn is reprinted with the permission of The Mathematical Association of America. It was originally published in the *American Mathematical Monthly,* Vol. 77, No. 6, June-July 1970, pp. 556-570.

Printed in the United States of America.

9 8 7 6 5 4 3 2 1

ISBN 0-387-90358-5   Springer-Verlag   New York
ISBN 3-540-90358-5   Springer-Verlag   Berlin   Heidelberg

Foreword

It is a privilege for me to write a foreword for this unusual book. The book is not primarily a reference work although many of the ideas and proofs are explained more clearly here than in any other source that I know. Nor is this a text of the customary sort. It is rather a record of a particular course and Gordon Whyburn's special method of teaching it. Perhaps the easiest way to describe the course and the method is to relate my own personal experience with a forerunner of this same course in the academic year 1937–1938.

At that time, the course was offered every other year with a following course in algebraic topology on alternate years. There were five of us enrolled, and on the average we knew less mathematics than is now routinely given in a junior course in analysis. Whyburn's purpose, as we learned, was to prepare us in minimal time for research in the areas in which he was interested. His method was remarkable.

It was not a lecture course. He gave definitions, drew pictures, gave examples, and stated theorems. Our task was to prove the theorems and to present the proofs in class. The ground rules were simple: we were each to work alone, and we were not to consult references until after the results in question had been proved in class. Moreover, no proof would be presented until each member of the class had either proved the results, or was willing to stop work on it.

If, as occasionally happened, none of us had proved one of the theorems, or one or more still wanted to work on a proof, Whyburn would simply give more theorems to prove, more definitions, and more examples. He agreed. that he would prove any result that we all gave up on, and though this never happened, I remember several times when theorems stayed unproved for more than two weeks.

Looking back, I marvel at the patience Whyburn displayed, listening gravely and thoughtfully to awkwardly constructed and badly presented proofs. He would point out gaps or errors in a proof without rancor or irritation. If he was ever bored or impatient with our efforts, it did not show.

The course was completely engrossing. We were far too immature mathematically to perceive sweeping patterns, but we could see the pattern within each group of the few theorems that we had before us at one time. We were utterly involved in learning how to construct proofs, how to construct mathematics. And we did get, by the end of year, to a stage where we could understand and hope to attack some of the interesting open problems in the area.

There are many advantages to organizing a course as Whyburn organized his. First, doing mathematics is just more fun than watching someone else do it, and students get involved. Second, an argument that one invents is almost unforgettable; it becomes a part of one's working equipment. An argument that one hears is much less likely to be remembered. And third, the method builds confidence. Constructing a proof for a good known theorem is next best to finding and proving a good theorem. It gives one hope of eventually creating mathematics.

There are disadvantages, or at least dangers in the method. Such a course cannot contain the detail and the organization that a polished set of good lectures might offer. It would just take too much time, and the primary goal of taking the student quickly to an area of research would be lost. What is needed, and what Whyburn accomplished, is to restrict the course to the very bones of the subject. Many of the theorems, and almost all of the ideas in the chosen mathematical area are here. Fleshing out these ideas is much simpler and can be done by the learner later.

There are other requirements, other than the "bare bones" mathematical structure, for a successful course of this sort. The course must begin mathematically where the students are. Whyburn had an instinctive grasp for the right place to begin.

The pacing of the course is also important. One will notice in this volume that the mathematical steps that the student is required to construct—the mathematical paces, if you like—are shorter in the earlier chapters. Near the beginning of the course many results are split into lemmas and preliminary propositions, so that the requisite arguments are within the abilities of the beginning student. Later in the course the student may find it necessary to invent these preliminary results for himself.

I believe that this book offers a stimulating and efficient path for a student to reach the important areas of research that concerned Gordon Whyburn just before his untimely death. The mathematical community is indebted to Professor Edwin Duda for completing and arranging the manuscript.

Berkeley, California   John L. Kelley
September 17, 1978

Preface

This book was prepared from a set of notes used by the late Professor Gordon T. Whyburn in an introductory course in topology. The intention of the book is to lead the student, through his own efforts, rather quickly to some important theorems concerning mappings on topological spaces, in particular mappings on continua or generalized continua. The importance of studying mappings and their behavior on topological spaces is treated in Whyburn's article entitled "Dynamic Topology," which appeared in 1970 in the American Mathematical Monthly and is reprinted in this volume.

The book, for the most part, is set up in the form of definitions and notations followed by exercises for the student to attempt and then solutions to the exercises. The student will develop and sharpen his ability to give proofs if he proceeds through the text in the fashion outlined in the note to the student.

The bibliography is meant to be historical, but we do not mean to imply that it is complete.

Coral Gables, Florida   Edwin Duda
July 4, 1978

Note to the Student

Read the material leading up to the exercises, and then attempt to develop your own proof for each exercise. The material is cumulative, as are some (but not all) of the exercises. For the proof of any given exercise you are permitted to use all the mathematics developed up to that point, and this includes all previous exercises proven or unproven.

Contents

PART A   1

Section I
Sets and Operations with Sets   3

Section II
Spaces   6

Section III
Directed Families   12

Section IV
Compact Sets and Bolzano-Weierstrass Sets   14

Section V
Functions   17

Section VI
Metric Spaces and a Metrization Theorem   22

Section VII
Diameters and Distances   28

Section VIII
Topological Limits   31

Section IX
Relativization   33

Section X
Connected Sets   34

Section XI
Connectedness of Limit Sets and Separations   38

Section XII
Continua   41

Section XIII
Irreducible Continua and a Reduction Theorem   43

Section XIV
Locally Connected Sets   45

Section XV
Property *S* and Uniformly Locally Connected Sets   48

Section XVI
Functions and Mappings   51

Section XVII
Complete Spaces   55

First Semester Examination   59

Section XVIII
Mapping Theorems   66

Section XIX
Simple Arcs and Simple Closed Curves   70

Section XX
Arcwise Connectedness   74

Appendix I
Localization of Property *S*   77

Appendix II
Cyclic Element Theory   79

PART B   83

Section I
Product Spaces   85

Section II
Decomposition Spaces 93

Section III
Component Decomposition 100

Section IV
Homotopy 105

Section V
Unicoherence 111

Section VI
Plane Topology 119

Appendix
Dynamic Topology 130
by G.T. Whyburn

Bibliography 145

Index 151

PART A

Sets and Operations with Sets

We shall somewhat imprecisely consider a set to be a collection of objects which satisfy a certain property. This is by no means a rigorous definition, and strictly speaking we shall not define the term set. We shall instead assume that the reader has an intuitive feeling for what constitutes a set and proceed accordingly.

If x is an element of a set A, then we denote this property by $x \in A$: If x is not an element of the set A, we write $x \notin A$. Suppose we have two sets A and B such that every element of the set A is also an element of the set B. Then we say that A is a subset of B and write $A \subset B$. A subset A of B is a proper subset if there exists an element $b \in B$ such that $b \notin A$. It is clear that $A = B$ provided $A \subset B$ and $B \subset A$.

By the union of two sets A and B we mean the set of all elements which are members of A and/or members of B, and we indicate this set by $A \cup B$, or by $A + B$. The intersection of two sets A and B is the set of all elements which are members of both A and B, and is represented by $A \cap B$, or by $A \cdot B$. If we consider $\cup$ and $\cap$ as operations on the sets A, B, and C, then these operations satisfy the distributive laws:

$$A \cup (B \cap C) = (A \cup B) \cap (A \cup C),$$
$$A \cap (B \cup C) = (A \cap B) \cup (A \cap C),$$

and the associative laws:

$$A \cup (B \cup C) = (A \cup B) \cup C,$$
$$A \cap (B \cap C) = (A \cap B) \cap C.$$

The complement of a set A relative to a set B is the set of elements $b \in B$ such that $b \notin A$, and we denote this set by $B \sim A$, or by $B - A$. If the set B

is clear in context and is fixed, then we refer to the complement of A relative to B as just the complement of A (A^c). We have the following laws:

$$(A \cup B)^c = A^c \cap B^c,$$
$$(A \cap B)^c = A^c \cup B^c.$$

The cartesian product of two sets A and B is the set of all 2-tuples (a,b) where $a \in A$ and $b \in B$.

The empty set or null set is denoted by $\varnothing$. We say that two sets A and B are *disjoint* if $A \cap B = \varnothing$.

A set whose elements can be put into a one-to-one correspondence with a subset of the positive integers is countable. If such a correspondence is established, and the resulting set is written in ascending order of its correspondence with the integers, then the arranged set is called a sequence.

EXERCISES I.

1. The union of the elements of any countable collection of countable sets results in a countable set.

2. The rational numbers are countable.

3. The real numbers and the complex numbers are not countable (uncountable).

SOLUTIONS.

1. The union of the elements of any countable collection of countable sets results in a countable set.

Let $\{E_n\}$ be a countable collection of sets such that each E_n has a countable number of elements; i.e., $E_n = \{a_{n1}, a_{n2}, a_{n3}, \ldots\}$ for each n. Then $E = \bigcup E_n = \bigcup_{m,n} a_{mn}$ is a countable set where the one to one correspondence with the integers is established in the following manner:

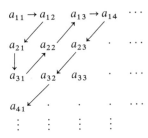

2. The rational numbers are countable.
Construct the sets E_n for each n by

$$E_1 = \{0,1\}, \qquad E_2 = \left\{\frac{1}{2}\right\},$$

$$E_3 = \left\{\frac{1}{3}, \frac{2}{3}\right\}, \qquad E_4 = \left\{\frac{1}{4}, \frac{2}{4}, \frac{3}{4}\right\}, \ldots,$$

$$E_n = \left\{\frac{1}{n}, \frac{2}{n}, \ldots, \frac{n-1}{n}\right\}, \ldots.$$

The set of all rationals between 0 and 1 is the union of the E_n's and so is countable. Similarly, the set of all rationals between n and $n+1$ is countable for every integer n; we denote this set by A_n. Then the set of all rationals $Q = \bigcup_n A_n$ is countable, since the set of all integers is countable.

3. The real numbers and the complex numbers are not countable (uncountable).
It is enough to show that the set of real numbers between 0 and 1 (denoted by $R_{[0,1]}$) is uncountable, since a subset of a countable set is countable. If $R_{[0,1]}$ is countable, then $R_{[0,1]} = \{a_n\}$. We can write each a_n in its decimal expansion: $a_n = .b_{n1}b_{n2}b_{n3} \ldots$. This expansion is unique if we require that if some a_n has a terminating decimal (i.e., $b_{nm} = 0$ for m sufficiently large) then a_n be written in the following form:

$$a_n = .b_{n1}b_{n2} \ldots b_{nN} = .b_{n1}b_{n2} \ldots (b_{nN} - 1)999 \ldots.$$

For example, $\frac{1}{4} = .25 = .24999 \ldots$. Now let r be a real number determined by the decimal $r = r_1 r_2 r_3 \ldots$ where $0 \neq r_n \neq b_{nn}$. Then $r \notin \{a_n\}$ but $r \in R_{[0,1]}$, which is a contradiction. Thus $R_{[0,1]}$ is uncountable.

SECTION II
Spaces

Definition. A *topological space* is a set X of elements together with a distinguished class of subsets called open sets satisfying the following axioms:

  (i) X and $\varnothing$ are open sets;
 (ii) arbitrary unions of open sets are open;
(iii) finite intersections of open sets are open.

EXAMPLE. The plane is a topological space if the open sets are the unions of interiors of circles.

Axiom iv (τ_1-Space Axiom). If x and y are points of X, $x \neq y$, then there exists an open set U_x containing x but not y.

Definition. A point p is said to be a *limit point* of a set M if and only if every open set containing p contains at least one point of M distinct from p.

Definition. A set A which contains all of its limit points is said to be *closed*.

Definition. If A is a set, then Cl(A) consisting of A and all of its limit points is called the *closure of A*.

Axiom v' (Hausdorff Axiom). If x and y are distinct points, then there exist disjoint open sets U_x and U_y such that $x \in U_x$, $y \in U_y$.

Axiom v (Regularity Axiom). If x is a point and E is a closed set not containing x, then there exist disjoint open sets U_x and U_E containing x and E respectively.

Axiom v'' (Normality Axiom). If A and B are disjoint closed sets, then there exist disjoint open sets U_A and U_B containing A and B respectively.

Definition. A *basis* for a topological space X is a collection $[G]$ of open sets such that every open set in X is a union of members of $[G]$.

Axiom vi (Countability Axiom). There exists a countable basis for the open sets in X, i.e., a countable collection R of open sets $R_1, R_2, \ldots$ such that every open set in X is a union of a subcollection of R.

Thus, if U is any open set in X and p is any point in U, there exists an integer m such that $p \in R_m \subset U$.

Such a space is called *perfectly separable*. A topological space which satisfies the τ_1-space axiom, Hausdorff axiom, regularity axiom, or normality axiom, is called respectively a τ_1-*space*, *Hausdorff space*, *regular space*, or *normal space*.

Definition. An infinite sequence of points $p_1, p_2, \ldots$ (not necessarily distinct) is said to *converge* to a point p, and we write $p_n \to p$, provided every open set about p contains all but a finite number of the p_i's.

Notes.
(1) If $p_n \to p$ in a Hausdorff space, then any subsequence $p_{n_i} \to p$.
(2) No sequence converges to more than one point if the space is Hausdorff.

EXERCISES II.

1. The closure of any set is closed.

2. If p is a limit point of $A \cup B$, then p is a limit point of at least one of the sets A and B.

3. If p is a limit point of a set M in a τ_1-space, then every open set containing p contains infinitely many distinct points of M.

4. A set is closed if and only if its complement is open.

5. The intersection of any collection of closed sets is a closed set.

6. The union of any finite collection of closed sets is closed.

7. Give an example of a non-Hausdorff τ_1-space.

8. Prove that a regular τ_1-space is a Hausdorff space.

9. Prove that a normal τ_1-space is regular.

10. If p is a point in a perfectly separable space X, then there exists a monotone decreasing sequence of open sets closing down on p; i.e., $Q_1 \supset Q_2 \supset \cdots$ is a sequence such that if U is any open set containing p, then $p \in Q_m \subset U$ for some integer m.

11. If X is τ_1 and perfectly separable, and p is a limit point of a set M, then M contains an infinite sequence of distinct points converging to p.

12. Lindelöf Theorem: Every collection $\mathscr{G}$ of open sets in a perfectly separable space contains a countable subcollection whose union is the same as the union of all of $\mathscr{G}$.

13. Tychonoff Lemma: Every regular, perfectly separable τ_1-space is normal.

Solutions.

1. The closure of any set is closed.

Let A be any set, $\text{Cl}(A)$ its closure, and p a limit point of $\text{Cl}(A)$. If U is any open set containing p, then U must contain at least one point of $\text{Cl}(A)$ distinct from p, say q. Then q is either a point of A or a limit point of A. If q is a limit point, since U is an open set about q, U must contain a point of A. Thus any open set about p contains a point of A, which implies that either $p \in A$ or p is a limit point of A. In either case $p \in \text{Cl}(A)$, so $\text{Cl}(A)$ is closed.

2. If p is a limit point of $A \cup B$, then p is a limit point of at least one of the sets A and B.

Suppose p is not a limit point of A or B. Then there exists an open set U containing p but no points of A distinct from p, and there exists an open set V such that V contains p and no points of B distinct from p. But this implies that $U \cap V$ is an open set containing p but no points of A or B distinct from p, which contradicts the assumption that p is a limit point of $A \cup B$.

3. If p is a limit point of a set M in a τ_1-space, then every open set containing p contains infinitely many distinct points of M.

Suppose an open set U_0 containing p contains only a finite number of points $p_1, p_2, \ldots, p_n$ of M. Since the space is τ_1, there exist open sets U_i such that $p_i \notin U_i$, $p \in U_i$. Let $V = \bigcap_0^n U_i$. Then $p \in V$, but no other point of M is contained in V, and p is a limit point of M. Contradiction.

4. A set is closed if and only if its complement is open.

Let A be a closed set. We want to show that A^c is open. It is enough to prove that for each $x \in A^c$, there exists an open set U_x containing x such that $U_x \subset A^c$, since then

$$A^c = \bigcup_{x \in A^c} U_x,$$

where each U_x is open. Thus A^c is the union of open sets and is then itself open.

Let $x \in A^c$. Then $x \notin A$, and x is not a limit point of A, since A contains all of its limit points. So there exists an open set U_x such that $x \in U_x$ but $U_x \cap A = \varnothing$ or $U_x \subset A^c$.

Let B be an open set, and x a limit point of B^c. Suppose $x \in B$. Then, since B is open, there exists a point $p \in B^c$ such that $p \in B$; but this is a contradiction, since $B \cap B^c = \varnothing$. Therefore $x \in B^c$, and B^c is closed.

5. The intersection of any collection of closed sets is a closed set.

Consider a collection of closed sets E_α, $\alpha \in A$. By DeMorgan's Law we know that

$$\left(\bigcap_{\alpha \in A} E_\alpha \right)^c = \bigcup_{\alpha \in A} E_\alpha^c.$$

Since E_α is closed, E_α^c is open. Hence

$$\left(\bigcap_{\alpha \in A} E_\alpha \right)^c = \bigcup_{\alpha \in A} E_\alpha^c \text{ is open,}$$

which implies $\bigcap_{\alpha \in A} E_\alpha$ is closed.

6. The union of any finite collection of closed sets is closed.

Let $\{E_i, i = 1, 2, \ldots, N\}$ be a finite collection of closed sets. Again by DeMorgan's Law,

$$\left(\bigcup_{i=1}^{N} E_i \right)^c = \bigcap_{i=1}^{N} E_i^c.$$

Since each E_i^c is open, the right side of the above identity is an open set. Hence $\bigcup_{i=1}^{N} E_i$ is closed.

7. Give an example of a non-Hausdorff τ_1-space.

Consider the set X of positive integers, and choose the open sets to be all subsets whose complements in X are finite and the null set. It is clear that X together with this collection of open sets satisfies the three axioms of a topological space.

Let i and j be positive integers and suppose $i < j$. If we consider the sets $U_i = \{1, 2, \ldots, j - 1, j + 1, \ldots\}$ and $U_j = \{1, 2, \ldots, i - 1, i + 1, \ldots\}$, then U_i and U_j are both open sets; $i \in U_i, j \notin U_i$; and $j \in U_j, i \notin U_j$. Thus X is a τ_1-space.

But i and j do not have disjoint neighborhoods. Suppose V_i and V_j are any two open sets containing i and j respectively. Since V_i and V_j have finite complements, there exist positive integers N_i and N_j such that if $n > N_i$, then $n \in V_i$, and if $n > N_j$, then $n \in V_j$. Thus, if $N = \max(N_i, N_j)$, then $n > N$ implies $n \in V_i$ and $n \in V_j$, so that $V_i \cap V_j \neq \varnothing$. Hence X cannot be Hausdorff.

8. Prove that a regular τ_1-space is a Hausdorff space.

We first note that in a τ_1-space X, every point is a closed set. Let $p \in X$, and consider the complement of p in X, $X \sim p$. Since X is a τ_1-space, for every $x \in X \sim p$ there exists an open set U_x containing x but not p; i.e., $U_x \subset X \sim p$. Thus $X \sim p$ is open, so $\{p\}$ must be a closed set.

If X is a regular τ_1-space, let x and y be distinct points in X. We know that x is a closed set in X, so by regularity, there exist disjoint open sets U_x and U_y containing x and y respectively. Hence X is a Hausdorff space.

9. Prove that a normal τ_1-space is regular.

Let X be a normal τ_1-space, $x \in X$, and E a closed set in X. Since X is τ_1, we know that $\{x\}$ is a closed set, and the normality of X then implies the existence of disjoint open sets U_x and U_E such that $x \in U_x$, $E \subset U_E$. Thus X must be regular.

10. If p is a point in a perfectly separable space X, then there exists a monotone decreasing sequence of open sets closing down on p; i.e., $Q_1 \supset Q_2 \supset \ldots$ is a sequence such that if U is any open set containing p, then $p \in Q_m \subset U$ for some integer m.

Since X is perfectly separable, there exists a countable basis $R = \{R_i\}_1^\infty$. Let $\{R_{n_i}\}_1^\infty$ be the subcollection of all elements of R which contain the point p. Set $Q_1 = R_{n_1}$, $Q_2 = Q_1 \cap R_{n_2}, \ldots, Q_k = Q_{k-1} \cap R_{n_k}, \ldots$. Then each Q_i is open, and since $Q_i = Q_{i-1} \cap R_{n_i}$, we see that $Q_i \subset Q_{i-1}$, so the Q_i's form a monotonic decreasing sequence of open sets.

Let U be an arbitrary open set containing p. Since X is perfectly separable, there exists an integer m such that $p \in R_m \subset U$. But then $R_m = R_{n_k}$ for some k, by construction, and thus $p \in Q_k = Q_{k-1} \cap R_{n_k} \subset R_{n_k} \subset U$.

11. If X is τ_1 and perfectly separable, and p is a limit point of a set M, then M contains an infinite sequence of distinct points converging to p.

Since X is perfectly separable, there exists a monotone decreasing sequence $Q_1, Q_2, \ldots$ of open sets closing down on p. Because p is a limit point of M, and Q_1 is an open set containing p, there exists point $p_1 \neq p$ such that $p_1 \in Q_1$, $p_1 \in M$.

Let U_1 be an open set such that $p \in U_1$, $p_1 \notin U_1$. Then $Q_2' = Q_2 \cap U_1$ is an open set containing p but not p_1. Thus Q_2' also contains a point $p_2 \in M$ such that $p \neq p_2 \neq p_1$.

Similarly, $Q_n' = Q_n \cap U_{n-1} \cap \cdots \cap U_1$ is an open set containing p, where U_i is an open set such that $p \in U_i$, $p_{i-1} \notin U_i$. Then Q_n' contains a point $p_n \in M$ such that $p \neq p_n \neq p_i$ for $i = 1, 2, \ldots, n-1$.

The sequence $p_1, p_2, \ldots$ has the desired properties. Clearly the p_i are distinct. Let V be any open set containing p. Then $Q_m \subset V$ for some integers m, and $V \supset Q_m \supset$ `, $Q_m' \supset Q_{m+1}' \supset \cdots$ implies that $\{p_m, p_{m+1}, \ldots\} \subset V$. Thus V contains almost all of the p_i, and $p_n \to p$.

12. Lindelöf Theorem: Every collection $\mathscr{G}$ of open sets in a perfectly separable space contains a countable subcollection whose union is the same as the union of all of $\mathscr{G}$.

Let $R_1, R_2, \ldots,$ be a countable basis for the space X. For each point p in an element $G \in \mathscr{G}$ there exists an integer m such that $p \in R_m \subset G$. Let $\{R_{n_i}\}$ be the collection of R_i's obtained for each such point p in an element $G \in \mathscr{G}$. For each i, select one $G \in \mathscr{G}$ with $R_{n_i} \subset G$, and call it G_i. Then $[G_i]$ is a countable subcollection of $\mathscr{G}$, and the union of the G_i's is the same set as the union of the elements of $\mathscr{G}$.

13. Every regular, perfectly separable τ_1-space is normal.

Let A and B be two disjoint closed sets in our space. By regularity, it follows that for each $x \in A$, there exists a neighborhood U of x such that $\mathrm{Cl}(U) \cap B = \varnothing$. Applying the Lindelöf Theorem to the collection $[U]$, we obtain a sequence of neighborhoods $U_1, U_2, \ldots$ covering A such that $\mathrm{Cl}(U_i) \cap B = \varnothing$ $(i = 1, 2, \ldots)$. Similarly, there exists a sequence of neighborhoods $V_1, V_2, \ldots$ covering B such that $\mathrm{Cl}(V_i) \cap A = \varnothing$ $(i = 1, 2, \ldots)$. Set

$$U_1^* = U_1, \qquad\qquad V_1^* = V_1 \sim V_1 \cap \mathrm{Cl}(U_1),$$
$$U_2^* = U_2 \sim U_2 \cap \mathrm{Cl}(V_1), \qquad V_2^* = V_2 \sim V_2 \cap (\mathrm{Cl}(U_1) \cup \mathrm{Cl}(U_2)),$$
$$\vdots \qquad\qquad\qquad\qquad \vdots$$
$$U_n^* = U_n \sim U_n \cap \left(\bigcup_1^{n-1} \mathrm{Cl}(V_i) \right), \quad V_n^* = V_n \sim V_n \cap \left(\bigcup_1^{n} \mathrm{Cl}(U_i) \right),$$
$$\vdots \qquad\qquad\qquad\qquad \vdots$$

Then for each n, U_n^* and V_n^* are open, since $U_n^* = (\bigcup_1^{n-1} \mathrm{Cl}(V_i))^c \cap U_n$ and $V_n^* = (\bigcup_1^{n} \mathrm{Cl}(U_i))^c \cap V_n$, the intersection of two open sets. Also, $U_n^* \cap A = U_n \cap A$ and $V_n^* \cap B = V_n \cap B$ because $\mathrm{Cl}(V_i) \cap A = \varnothing$ and $\mathrm{Cl}(U_i) \cap B = \varnothing$. Let $U^* = \bigcup_1^{\infty} U_i^*$ and $V^* = \bigcup_1^{\infty} V_i^*$. Then U^* and V^* are open and contain A and B respectively. Also $U^* \cap V^* = \varnothing$. For if not, then there would exist integers m and n such that $U_m^* \cap V_n^* \neq \varnothing$. But this is not possible, by the construction of the U_i^*'s and the V_i^*'s. Thus the space is normal.

SECTION III

Directed Families

Definition. A nonempty collection $\mathscr{F}$ of nonempty sets is called a *directed family* provided $F_1, F_2 \in \mathscr{F}$ imply that $F_1 \cap F_2$ contains at least one element F_3 of $\mathscr{F}$.

EXAMPLE. Consider a sequence $p_1, p_2, \ldots$. Let $F_n = \{p_n, p_{n+1}, \ldots\}$. Then $\mathscr{F} = \{F_n, n = 1, 2, \ldots\}$ is directed family. In this sense, a directed family is an extension of the notion of sequence.

Definition. A point p is called a *cluster point* of a directed family provided every open set about p intersects each element F of the family.

Definition. A directed family $\mathscr{F}$ *converges* to a point p if and only if every open set about p contains some element of the family.

Definition. If $\mathscr{E}$ and $\mathscr{F}$ are directed families, then $\mathscr{E}$ is a (directed) *underfamily* of $\mathscr{F}$ provided each element of $\mathscr{F}$ contains some element of $\mathscr{E}$.

Note. If $\mathscr{F}$ converges to p, then so does every underfamily.

EXERCISES III.

1. A point p is a cluster point of a directed family $\mathscr{F}$ provided some underfamily of $\mathscr{F}$ converges to p.

2. A topological space X is a Hausdorff space if and only if each directed family of sets in X converges to at most one point in X.

3. If $\mathscr{F}$ converges to p and X is a Hausdorff space, then no other point of X is a cluster point of $\mathscr{F}$.

SOLUTIONS.

1. A point p is a cluster point of a directed family $\mathscr{F}$ provided some underfamily of $\mathscr{F}$ converges to p.

Suppose that p is a cluster point of a directed family $\mathscr{F} = \{F_\beta : \beta \in B\}$. Consider the set $\{U_\alpha : \alpha \in A\}$ of all open sets in X containing p. Let $\mathscr{E} = \{U_\alpha \cap F_\beta : \alpha \in A, \beta \in B\}$. Each element of $\mathscr{E}$ is nonempty, since p is a cluster point of $\mathscr{F}$. Also, $(U_{\alpha_1} \cap F_{\beta_1}) \cap (U_{\alpha_2} \cap F_{\beta_2}) \supset (U_{\alpha_1} \cap U_{\alpha_2}) \cap F_{\beta_3} \in \mathscr{E}$, because $\mathscr{F}$ is a directed family and $U_{\alpha_1} \cap U_{\alpha_2}$ is an open set containing p. Hence $\mathscr{E}$ is a directed family and clearly an underfamily of $\mathscr{F}$. Let V be any open set containing p. Then $V = U_\alpha$ for some α, and $V = U_\alpha \supset U_\alpha \cap F_\beta$ for any $\beta \in B$, so $\mathscr{E}$ must converge to p.

Now let $\mathscr{F}$ be a directed family, $\mathscr{F} = \{F_\alpha : \alpha \in A\}$, and $\mathscr{E}$ an underfamily of $\mathscr{F}$ which converges to p, $\mathscr{E} = \{E_\beta : \beta \in B\}$. Suppose U is an arbitrary open set about p, and F_α is some element of $\mathscr{F}$. Since $\mathscr{E}$ converges to p, U contains some E_β; and since $\mathscr{E}$ is an underfamily of $\mathscr{F}$, F_α contains some E_α. But $E_\beta \cap E_\alpha \supset E_\gamma \in \mathscr{E}$, where $E_\gamma \subset E_\beta \subset U$ and $E_\gamma \subset E_\alpha \subset F_\alpha$. Thus $U \cap F_\alpha \supset E_\gamma \neq \varnothing$, and $\mathscr{F}$ has p as a cluster point.

2. A topological space X is a Hausdorff space if and only if each directed family of sets in X converges to at most one point in X.

Suppose X is a Hausdorff space and $\mathscr{F} = \{F_\alpha\}$ is a directed family in X which converges to points p and q in X. Then every open set containing p contains some F_α, and every open set containing q contains some F_β. Let U_p and U_q be disjoint open sets in X containing p and q respectively. Then $U_p \supset F_\alpha$, $U_q \supset F_\beta$ for some α and β, and $F_\alpha \cap F_\beta = \varnothing$; but this contradicts the assumption that $\mathscr{F}$ is a directed family, so $p = q$.

Assume that every directed family of sets in X converges to at most one point. Let x and y be distinct points in X, and suppose that for every open set M_α containing x and every open set N_β containing y, $M_\alpha \cap N_\beta \neq \varnothing$. Then we can construct a directed family $\mathscr{F}$ where $\mathscr{F} = \{M_\alpha \cap N_\beta : M_\alpha$ is an open set containing x, N_β is an open set containing $y\}$. But then every open set containing x contains a member of $\mathscr{F}$, and the same holds true for open sets containing y, so that $\mathscr{F}$ converges to both x and y, which contradicts our assumption. Thus X is Hausdorff.

3. If $\mathscr{F}$ converges to a point p and X is a Hausdorff space, then no other point of X is a cluster point of $\mathscr{F}$.

We know that if $\mathscr{F}$ converges to p, then every underfamily of $\mathscr{F}$ converges to p, and since X is Hausdorff, every underfamily converges only to p. Suppose there existed a point $q \neq p$ such that q was a cluster point of $\mathscr{F}$. Then some underfamily of $\mathscr{F}$ must converge to q, which is a contradiction, and the result follows.

SECTION IV

Compact Sets and Bolzano–Weierstrass Sets

Definition. A set K in a topological space X is said to be *compact* if and only if any collection $[G]$ of open sets covering K (i.e., $\bigcup G \supset K$) has a finite subcollection also covering K.

Definition. A set H is called a *Bolzano–Weierstrass (B-W)* set provided every infinite subset of H has at least one limit point in H.

Definition. A set M is *conditionally compact* provided every infinite subset of M has at least one limit point, but not necessarily in M.

EXERCISES IV.

1. In a Hausdorff space, every compact set is closed.

2. If $K_1 \supset K_2 \supset K_3 \supset \cdots$ is a monotone decreasing sequence of nonempty compact sets in a Hausdorff space, then $\bigcap_1^\infty K_n$ is also nonempty.

3. A set K is compact if and only if each directed family of sets in K has at least one cluster point in K.

4. Every compact set is a B-W set.

5. In a perfectly separable Hausdorff space, every B-W set is closed.

6. If $K_1 \supset K_2 \supset \cdots$ is a monotone decreasing sequence of closed nonempty B-W sets in a perfectly separable τ_1-space, then $\bigcap_1^\infty K_n$ is nonempty.

7. Borel Theorem: Let X be τ_1 and perfectly separable. Then every B-W set in X is compact.

8. If a set M is conditionally compact and $M \subset X$, where X is τ_1, regular, and perfectly separable, then $\text{Cl}(M)$ is compact.

9. Every compact Hausdorff space is normal.

14

SOLUTIONS.

1. In a Hausdorff space, every compact set is closed.

Suppose X is Hausdorff, M is a compact set in X, and p is a limit point of M. Let x be any point of M distinct from p. There exist disjoint open sets U_x and V_x containing x and p respectively. If $p \notin M$, then $[U_x]$ for $x \in M$ is an open cover for M and there exists a finite subcover $U_1, U_2, \ldots, U_n$ for M. Then the intersection of the corresponding open sets $V_1, V_2, \ldots, V_n$ is an open set containing p but no other points of M. But p was a limit point, so p must be contained in M.

2. If $K_1 \supset K_2 \supset K_3 \supset \cdots$ is a monotone decreasing sequence of nonempty compact sets in a Hausdorff space, then $\bigcap_1^\infty K_n$ is also nonempty.

Since X is Hausdorff, each K_i is closed, so each K_i^c is open. Suppose K_1 has no point in every K_i $(i \geq 1)$. Then $[K_i^c]$ is an open cover for K_1, and it can be reduced to a finite subcover $K_{m_1}^c, K_{m_2}^c, \ldots, K_{m_n}^c$; i.e.,

$$K_1 \subset K_{m_1}^c \cup \cdots \cup K_{m_n}^c = \left(\bigcap_1^n K_{m_i} \right)^c.$$

Then no point of K_1 lies in the intersection $\bigcap_1^n K_{m_i} = K_{m_n}$. But this implies $K_{m_n} = \varnothing$, which is a contradiction. Thus $\bigcap_1^\infty K_n \neq \varnothing$.

3. A set K is compact if and only if each directed family of sets in K has at least one cluster point in K.

Let K be compact and $\mathscr{F}$ a directed family of sets in K with no cluster point in K. If $p \in K$, then there exists a neighborhood N_p about p that fails to intersect some $F_\alpha \in \mathscr{F}$. The collection of all such neighborhoods forms an open covering for K. Hence there exists a finite subcovering $N_1, N_2, \ldots, N_n$ for K such that $N_i \cap F_{\alpha_i} = \varnothing$ for some $F_{\alpha_i} \in \mathscr{F}$. Let $A = \bigcap_1^n F_{\alpha_i}$. Then A must contain another member of $\mathscr{F}$, say F_β. But the intersection of F_β with the open cover of K is empty. Thus $F_\beta = \varnothing$ or $F_\beta \subset K^c$, both of which are contradictions. Thus $\mathscr{F}$ must have a cluster point in K.

Now assume that every directed family of sets in K has at least one cluster point in K, and suppose K is not compact. Then there exists an open cover $\{F_\alpha : \alpha \in A\}$ of K which cannot be reduced to a finite subcover. Consider the sets $G_\alpha = F_\alpha^c \cap K$. Using DeMorgan's Laws, we see that all finite intersections of G_α's are nonempty, and the sets G_α together with their finite intersections form a directed family $\mathscr{G}$. Then $\mathscr{G}$ has at least one cluster point p in K. It is clear that another characterization of a cluster point p is that $p \in \bigcap_\alpha \mathrm{Cl}(G_\alpha)$, so that $\bigcap_\alpha \mathrm{Cl}(G_\alpha) \neq \varnothing$. But then it follows that $\bigcap_\alpha G_\alpha \neq \varnothing$, since G_α contains all of its limit points, which lie in K by construction, and $p \in K$. Thus $\varnothing \neq \bigcap_\alpha G_\alpha = \bigcap_\alpha (F_\alpha^c \cap K) = K \bigcap_\alpha (\bigcup F_\alpha)^c$, which implies that $\{F_\alpha : \alpha \in A\}$ is not a cover for K. Hence K must be compact.

4. Every compact set is a B-W set.

Let K be a compact set and E an infinite subset of K. Suppose E has no limit point in K. Then corresponding to each $x \in K$ there is an open set U_x containing x such that U_x contains no points of E distinct from x. The set of all such U_x for $x \in K$ is an open cover of K, so there exists a finite subcover $U_1, U_2, \ldots, U_n$. Each U_i contains at most one point of E, so E can have at most n points. This is a contradiction, and therefore E has at least one limit point in K.

5. In a perfectly separable Hausdorff space, every B-W set is closed.

Let H be a B-W set in a perfectly separable Hausdorff space, and let p be a limit point of H. Then H contains a sequence of distinct points $p_1, p_2, \ldots$ such that $p_n \to p$.

The set $\{p_1, p_2, \ldots\}$ is infinite and must have a limit point q in H. Suppose $q \neq p$. Let U_p and U_q be disjoint open sets containing p and q. Since $p_n \to p$, U_p contains all but a finite number of the p_i, which implies that U_q can contain at most a finite number. But since the space was Hausdorff, U_q must contain an infinite number of points. Thus $p = q$ and $p \in H$.

6. If $K_1 \supset K_2 \supset \cdots$ is a monotone decreasing sequence of closed nonempty B-W sets in a perfectly separable τ_1-space, then $\bigcap_1^\infty K_n$ is nonempty.

For each n, choose points $p_n \in K_n$. If some p_m is chosen an infinite number of times, then $p_m \in K_n$ for each n, because $K_n \supset K_{n+1}$, so that $p_m \in \bigcap_1^\infty K_n$. Suppose no p_m appears an infinite number of times. Then the set $[p_n]$ is an infinite subset of K_1, and hence it has a limit point p in K_1. But p is a limit point of the subset $\{p_n, p_{n+1}, \ldots\}$ of K_n for every n, and since K_n is closed, we have $p \in K_n$ for each n. Thus $p \in \bigcap_1^\infty K_n \neq \varnothing$.

7. Borel Theorem: Let X be τ_1 and perfectly separable. Then every B-W set in X is compact.

Consider the B-W set K, and suppose K is not compact. Then there exists an open covering $[G_\alpha]$ of K which has no finite subcovering. However, by the Lindelöf Theorem, there exists a countable subcovering $[G_n]$ of K. Construct a sequence p_n by choosing $p_n \in K \sim (U_1^n G_i)$. If no finite subcollection of $[G_n]$ covers K, then p_n is an infinite sequence. If p_n is an infinite sequence it has a limit point in K. Since $p \in k$, it follows that some G_n contains p and hence infinitely many of the p_n. This is impossible by the construction of the sequence and therefore some finite subcollection of $[G_n]$ covers K and K is compact.

8. If a set M is conditionally compact and $M \subset X$, where X is τ_1, regular, and perfectly separable, then $\mathrm{Cl}(M)$ is compact.

Since X is regular and perfectly separable, $\mathrm{Cl}(M)$ is compact provided $\mathrm{Cl}(M)$ is a B-W set. We shall show the $\mathrm{Cl}(M)$ is a B-W set. Suppose not. Then $\mathrm{Cl}(M)$ contains an infinite subset E which has no limit point in $\mathrm{Cl}(M)$. Since $\mathrm{Cl}(M)$ is closed, E has no limit point at all, and hence E is closed in X. By regularity, to each $x \in X$ there corresponds an open set V_x containing x such that $\mathrm{Cl}(V_x)$ contains no points of E distinct from x. In other words, $\mathrm{Cl}(V_x)$ contains at most one point of E. Then $[V_x]$ is an open cover for X, and by the Lindelöf Theorem, this can be reduced to a countable covering $[V_n]$. We know that for each N, $\bigcup_1^N V_n$ contains at most N points of E. Since E is infinite and hence M is also, we can pick a sequence $[p_n]$ of distinct points of M such that $p_N \notin \bigcup_1^N V_n$. Thus $[p_n]$ is an infinite subset of a conditionally compact set M, so that $[p_n]$ has a limit point p. For some K, $p \in V_K$ but V_K is an open set containing p and at most K of the points p_n. Thus p can't be a limit point of $[p_n]$, and M is a B-W set and hence compact.

9. Every compact Haudorff space is normal.

Suppose A and B are disjoint sets. Choose an arbitrary point $a \in A$. Then for each $b \in B$, there exist disjoint open sets $U_{a,b}$ and $V_{a,b}$ containing a and b respectively. Since $[V_{a,b}]$ is an open cover for B and B is compact, there exists a finite subcover $V_{a,1}$, $V_{a,2}, \ldots, V_{a,N}$. Then $U_{a,B} = U_{a,1} \cap \cdots \cap U_{a,N}$ and $V_{a,B} = V_{a,1} \cup \cdots \cup V_{a,N}$ are disjoint open sets containing a and B. Covering A and B with sets of the form $U_{a,B}$ and $V_{a,B}$ and then using the compactness of A we obtain, in a manner similar to that above, disjoint open sets $U_{A,B}$ and $V_{A,B}$ containing A and B respectively. Thus the space is normal.

Functions

Definition. If X and Y are two topological spaces, a *function* f from X to Y is any law which assigns to each element of X a unique element of Y. The action of this law is represented by $f(x) = y$, where $x \in X$, $y \in Y$.

A function f from X to Y ($f: X \to Y$) is said to be *onto* Y if, for each $y \in Y$, there exists an $x \in X$ such that $f(x) = y$. The *range* of f (or the *image* of f), $f(X)$, is the set of all $y \in Y$ such that $y = f(x)$ for some $x \in X$. Thus f is onto Y provided $f(X) = Y$.

A function f is *one-to-one* (1-1) provided $f(x) = f(y)$ implies $x = y$. If f is 1-1, then we can define a function f^{-1} (called f-inverse) on the range of f by $f^{-1}(y) = x$ if $f(x) = y$. Even if f is not 1-1, we can still define $f^{-1}(y)$, but f^{-1} need not be a function in this case. In general, $f^{-1}(y) = \{x \in X : f(x) = y\}$.

Definition. A function $f: X \to Y$, X and Y topological spaces, is *continuous at a point* $p \in X$ if for any open set V about $f(p)$, there exists an open set U about p such that $f(U) \subset V$. If f is continuous at all points $x \in X$, then f is said to be a *continuous function* or *mapping*.

EXERCISES V.

1. (a) For any directed family $\mathscr{M}$ in X, the collection $\mathscr{N} = [f(M)]$, $M \in \mathscr{M}$, is a directed family in Y.
   (b) For any directed family $\mathscr{N}$ in $f(X)$, the collection $\mathscr{M} = [f^{-1}(N)]$, $N \in \mathscr{N}$, is a directed family in X.
   (c) Given $\mathscr{M}$ in X and $\mathscr{N}'$ an underfamily of $\mathscr{N} = f(\mathscr{M})$, $\mathscr{M}$ and $f^{-1}(\mathscr{N}')$ are "related" in the sense that they have a common underfamily.

2. A function $f: X \to Y$ is continuous if and only if the inverse of every open (closed) set in Y is open (closed) in X.

3. A function $f: X \to Y$ is continuous at $p \in X$ if and only if for any directed family $\mathscr{F}$ in X converging to p, $f(\mathscr{F})$ converges to $f(p)$.

4. If Y is the real or complex number system with the usual topology, a function $f: X \to Y$ is continuous at a point $p \in X$ provided that for each $\varepsilon > 0$ there exists an open set U in X about p such that $|f(x) - f(p)| < \varepsilon$ for $x \in U$.

5. Suppose that X is τ_1 and perfectly separable. Then a function $f: X \to Y$ is continuous at $p \in X$ if and only if for any sequence $x_1, x_2, \ldots$ in X converging to p, we have $f(x_1), f(x_2), \ldots$ converging in Y to $f(p)$.

6. *Urysohn's Lemma*: If A and B are any nonempty disjoint closed sets in a normal space X, then there exists a continuous function $f: X \to [0,1]$ such that $f(A) = 0$ and $f(B) = 1$.

SOLUTIONS.

1. (a) For any directed family $\mathscr{M}$ in X, the collection $\mathscr{N} = [f(M)]$, $M \in \mathscr{M}$, is a directed family in Y.

Let $M_\alpha, M_\beta \in \mathscr{M}$. Then $f(M_\alpha) = \{f(x) \in Y : x \in M_\alpha\}$, and $f(M_\beta)$ is defined similarly. Thus $f(M_\alpha) \cap f(M_\beta) = \{f(x) : x \in M_\alpha \cap M_\beta\}$. Since $\mathscr{M}$ is a directed family, $M_\alpha \cap M_\beta \supset M_\gamma$ for $M_\gamma \in \mathscr{M}$, so we have $f(M_\alpha) \cap f(M_\beta) \supset f(M_\gamma)$, and $\mathscr{N} = [f(M)]$, $M \in \mathscr{M}$, is a directed family.

(b) For any directed family $\mathscr{N}$ in $f(X)$, the collection $\mathscr{M} = [f^{-1}(N)]$, $N \in \mathscr{N}$, is a directed family in X.

Let $M_\alpha = f^{-1}(N_\alpha)$, $M_\beta = f^{-1}(N_\beta)$ be elements of $\mathscr{M}$. Then $M_\alpha \cap M_\beta = \{x : f(x) \in N_\alpha \cap N_\beta\}$. Since $\mathscr{N}$ is a directed family, $N_\alpha \cap N_\beta \supset N_\gamma \in \mathscr{N}$. Then $M_\alpha \cap M_\beta \supset \{x : f(x) \in N_\gamma\} = M_\gamma$, and $\mathscr{M}$ is a directed family in X.

(c) Given $\mathscr{M}$ in X and $\mathscr{N}'$ an underfamily of $\mathscr{N} = f(\mathscr{M})$, $\mathscr{M}$ and $f^{-1}(\mathscr{N}')$ are "related" in the sense that they have a common underfamily.

Consider the directed family

$$\mathscr{M}' = \{M_\alpha \cap f^{-1}(N'_\beta) : M_\alpha \in \mathscr{M}, N'_\beta \in \mathscr{N}'\}.$$

Clearly each set in $\mathscr{M}$ and in $f^{-1}(\mathscr{N}')$ contains a set of $\mathscr{M}'$, and each element of $\mathscr{M}'$ is nonempty. Assuming the opposite, then $M_\alpha \cap f^{-1}(N'_\beta) = \varnothing$ and $f(M_\alpha) \cap N'_\beta = \varnothing$, since $f^{-1}(N'_\beta)$ is an inverse set. However, $f(M_\alpha) = N_\alpha$ contains N'_α of $\mathscr{N}'$ if $\mathscr{N}'$ is an underfamily of $\mathscr{N}$, and we cannot have $N'_\alpha \cap N'_\beta = \varnothing$. Finally, we verify that the intersection of any two elements of $\mathscr{M}'$ contains a third element of $\mathscr{M}'$, since

$$M_\alpha \cap f^{-1}(N'_\beta) \cap M_\gamma \cap f^{-1}(N'_\delta) \supset M_\varepsilon \cap f^{-1}(N'_p)$$

if M_ε lies in $M_\alpha \cap M_\gamma$ and N'_p lies in $N'_\beta \cap N'_\delta$.

2. A function $f : X \to Y$ is continuous if and only if the inverse of every open (closed) set in Y is open (closed) in X.

Assume that f is continuous and V is any open set in Y. If $V \cap f(X) = \varnothing$, then $f^{-1}(V) = \varnothing$, which is open, so we can assume that $V \cap f(X) \neq \varnothing$ and $f^{-1}(V) \neq \varnothing$. Let $p \in f^{-1}(V)$. Then $f(p) \in V$, so there exists an open set U containing p such that $f(U) \subset V$. But then $U \subset f^{-1}(V)$. Thus $f^{-1}(V)$ is open.

Suppose the inverse of every open set in Y is open in X. Let p be a point in X and V an open set containing $f(p)$. Then $f^{-1}(V) = U$ is open in X and contains p, and $f(U) = ff^{-1}(V) \subset V$. Hence f is continuous, since p is arbitrary.

The dual statement for closed sets is proved by taking complements.

3. A function $f : X \to Y$ is continuous at $p \in X$ if and only if for any directed family $\mathscr{F}$ in X converging to p, $f(\mathscr{F})$ converges to $f(p)$.

Consider the function $f : X \to Y$ which is continuous at p, and $\mathscr{F}$ a directed family in X converging to p. We know that $f(\mathscr{F})$ is a directed family in Y. Let V be an open set containing $f(p)$. There exists an open set U containing p such that $f(U) \subset V$. Since $\mathscr{F}$ converges to p, $F_\alpha \subset U$ for some $F_\alpha \in \mathscr{F}$. Then $f(F_\alpha) \subset V$, and $f(\mathscr{F})$ converges to $f(p)$.

Suppose that for every directed family $\mathscr{F}$ in X converging to p, $f(\mathscr{F})$ converges to $f(p)$. Consider the directed family $\mathscr{F}$ consisting of all open sets in X containing p. Let V be any open set containing $f(p)$. Since $f(\mathscr{F})$ converges to $f(p)$, $f(F_\alpha) \subset V$ for some $F_\alpha \in \mathscr{F}$. Then F_α is the required open set in X, and f is continuous at p.

4. If Y is the real or complex number system with the usual topology, a function f: $X \rightarrow Y$ is continuous at a point $p \in X$ provided that for each $\varepsilon > 0$ there exists an open set U in X about p such that $|f(x) - f(p)| < \varepsilon$ for $x \in U$.

Assume that f is continuous at p, and let $\varepsilon > 0$ be given. Then $V = \{ y \in Y : |y - f(p)| < \varepsilon \}$ is an open set containing $f(p)$. There exists an open set U containing p in X such that $f(U) \subset V$; i.e.,

$$|f(x) - f(p)| < \varepsilon \quad \text{for } x \in U.$$

Suppose for each $\varepsilon > 0$ there exists an open set U such that $|f(x) - f(p)| < \varepsilon$ for $x \in U$. Let V be an open set containing $f(p)$. Since in the usual topology interiors of circles form a basis, $f(p) \in V_\varepsilon(q) = \{ y \in Y : |y - q| < \varepsilon' \} \subset V$; i.e., $f(p)$ is contained in the interior of a circle of radius ε' about some point q. Then $f(p) \in V_{\varepsilon''}(q) \subset V_\varepsilon(q)$. By our assumption there exists an open set U containing p such that $|f(x) - f(p)| < \varepsilon''$ for all $x \in U$. Hence $f(U) \subset V_{\varepsilon''}(f(p)) \subset V$, and f is continuous at p.

5. Suppose that X is τ_1 and perfectly separable. Then a function $f: X \rightarrow Y$ is continuous at $p \in X$ if and only if for any sequence $x_1, x_2, \ldots$ in X converging to p, we have $f(x_1)$, $f(x_2), \ldots$ converging in Y to $f(p)$.

Assume that f is continuous at $p \in X$ and $\{x_n\}$ converges to p. If V is an open set containing $f(p)$, there exists an open set U containing p such that $f(U) \subset V$. Since $x_n \rightarrow p$, $x_n \in U$ for all n larger than some integer N. But then $f(x_n) \in V$ for $n > N$, and $f(x_n) \rightarrow f(p)$.

Suppose that for any sequence $\{x_n\}$ in X converging to a point p, $f(x_n)$ converges to $f(p)$. If f were not continuous at p, there would be an open set V about $f(p)$ such that, for each open set U containing p, there exists some $x \in U$ such that $f(x) \notin V$. Since X is perfectly separable, there is a sequence of open sets closing down on p. We can construct a sequence of distinct points $\{x_i\}$ (since X is τ_1) where $x_n \in Q_n$ and $f(x_n) \notin V$. By our construction, $x_n \rightarrow p$ but $\{f(x_n)\}$ does not converge to $f(p)$. Thus we have a contradiction, and f must be continuous at p.

6. *Urysohn's Lemma*: If A and B are nonempty disjoint closed sets in a normal space X, then there exists a continuous function $f: X \rightarrow [0,1]$ such that $f(A) = 0$ and $f(B) = 1$.

Since X is normal, there exist two disjoint open sets $U_{1/2}$ and $V_{1/2}$ containing A and B, or equivalently,

$$A \subset U_{1/2} \subset V_{1/2}^c \subset B^c.$$

The two sets A and $U_{1/2}^c$ and closed and disjoint, so again by normality there are disjoint open sets $U_{1/4}$ and $V_{1/4}$ containing A and $U_{1/2}^c$ respectively, and

$$A \subset U_{1/4} \subset V_{1/4}^c \subset U_{1/2}.$$

Similarly there exist disjoint open sets $U_{3/4}$ and $V_{3/4}$ containing $V_{1/2}^c$ and B, and

$$V_{1/2}^c \subset U_{3/4} \subset V_{3/4}^c \subset B^c.$$

Combining the above chains, we have

$$A \subset U_{1/4} \subset V_{1/4}^c \subset U_{1/2} \subset V_{1/2}^c \subset U_{3/4} \subset V_{3/4}^c \subset B^c.$$

We can further extend this chain by induction: for any integer m there is a chain

$$A \subset U_{1/2^m} \subset V_{1/2^m}^c \subset U_{2/2^m} \subset V_{2/2^m}^c \subset \cdots \subset U_{(2^m-1)/2^m} \subset V_{(2^m-1)/2^m}^c \subset B^c,$$

where $U_{k/2^m}$ and $V_{k/2^m}$ are open sets for each integer k, $1 \leq k < 2^m$. The construction of this chain results in the following properties:

(i) for each dyadic rational in $[0,1]$, $r = k/2^m$, k and m integers, there exist open sets U_r and V_r such that

$$A \subset U_r \subset V_r^c \subset B^c;$$

(ii) for any two dyadic rationals $r_1 < r_2$ we have

$$U_{r_1} \subset V_{r_1}^c \subset U_{r_2} \subset V_{r_2}^c.$$

Henceforth r and r_1 will denote dyadic rationals.

We define our function $f: X \to [0,1]$ by

$$f(x) = \begin{cases} \text{g.l.b.}\{r : x \in U_r\} & \text{if } x \in \bigcup U_r, \\ 1 & \text{if } x \notin \bigcup U_r. \end{cases}$$

By our construction, if $x \in A$, then $x \in U_r$ for every dyadic rational r, and if $x \in B$, then $x \notin U_r$ for any r. Thus $f(A) = 0$ and $f(B) = 1$.

To complete the proof we need only show that f is continuous. It is enough to show that $f^{-1}(B)$ is open for B an arbitrary member of a basis $\mathcal{B}$ for the topology of $[0,1]$. Since we are assuming the usual topology for $[0,1]$, one such basis is

$$\{[0,a), (b,1], (c,d) : a, b, c, d \text{ are irrational}\}.$$

Since $f^{-1}(c,d) = f^{-1}([0,d) \cap (c,1]) = f^{-1}[0,d) \cap f^{-1}(c,1]$, we need only show that $f^{-1}[0,a)$ and $f^{-1}(b,1]$ are open for each irrational a and b. But $f^{-1}[0,a) = \bigcup_{r<a} U_r$ and $f^{-1}(b,1] = \bigcup_{b<r} V_r$, so both of these sets are open and f is continuous.

SECTION VI

Metric Spaces and a Metrization Theorem

Definition. A *distance function* ρ in a set X is a nonnegative real-valued function defined for each pair of points $x, y \in X$ and satisfying:

(i) $\rho(x,y) = 0$ if and only if $x = y$,

(ii) $\rho(x,y) = \rho(y,x)$,

(iii) $\rho(x,z) \le \rho(x,y) + \rho(y,z)$ (triangle inequality).

Definition. A set X with a distance function ρ in which a set U is open provided each point $x \in U$ lies in an "open" sphere $V_r(x)$ contained in U is called a *metric space*. (For any real number $r > 0$, the "*open*" *sphere* $V_r(x)$ consists of all points p of X satisfying $\rho(x,p) < r$.)

Definition. A topological space is *metrizable* provided it admits a distance function ρ whose topology is equivalent to the given topology in X, i.e., so that a set U is open if and only if it is empty or is the union of "open" spheres given by ρ.

Equivalently, a point p is a limit point of a set M if and only if for every $\varepsilon > 0$, there exists a point x of M with $0 < \rho(x,p) < \varepsilon$.

Definition. A space or a set M is *separable* provided it contains a countable dense subset; i.e., a countable set P such that $\mathrm{Cl}(P) \supset M$.

EXAMPLE.

1. The real number system $\mathbb{R}$ and the complex number system $\mathbb{C}$ with metric $\rho(x, y) = |x - y|$.

22

2. The n-dimensional Euclidean plane E^n with distance function

$$\rho(x, y) = \sqrt{\sum_1^n (x_i - y_i)^2}.$$

3. The Hilbert space l_2, the points being sequences of real numbers $(x_1, x_2, \ldots)$, where $\sum_1^\infty x_i^2 < \infty$ and distance $\rho(x,y) = \sqrt{\sum_1^\infty (x_i - y_i)^2}$.

4. The Hilbert parallelotrope Q_ω (subset of l_2), consisting of sequences $(x_1, x_2, \ldots)$ where $0 \leq x_i \leq 1/i$ are points and with the same distance function $\sqrt{\sum_1^\infty (x_i - y_i)^2}$.

5. Q'_ω, consisting of sequences of reals $(x_1, x_2, \ldots)$ with $0 \leq x_i \leq 1$ as points and the distance defined by

$$\rho(x, y) = \sum_1^\infty \frac{|x_i - y_i|}{2^i}.$$

6. The Cartesian product $X \times Y$ of two metric spaces X and Y (with distance functions ρ_X and ρ_Y) consisting of pairs (x, y), $x \in X$, $y \in Y$, as points, where the distance function is defined by

$$\rho(x, y) = \sqrt{\rho_x(x_1, y_1)^2 + \rho_y(x_2, y_2)^2}$$

for $x = (x_1, x_2)$ and $y = (y_1, y_2)$.

EXERCISES VI.

1. Metrization Theorem: Every regular, perfectly separable τ_1-space is metrizable.

2. Every separable metric space satisfies Axioms (i)–(vi).

3. Show that the distance functions in Examples 2, 3, 5, and 6 satisfy the triangle inequality.

SOLUTIONS.

1. Metrization Theorem: Every regular, perfectly separable τ_1-space is metrizable.

Suppose that X is a perfectly separable, regular τ_1-space, and let $\{R_1, R_2, \ldots\}$ be a basis for X. By the Tychonoff Lemma, X is normal. Consider all pairs of basic elements R_i, R_j such that $\mathrm{Cl}(R_i) \subset R_j$. Since the collection of all such pairs is countable, it can be arranged in a sequence $P_1, P_2, \ldots$. By Urysohn's Lemma, corresponding to each pair $P_n = R_i$, R_j there is a continuous function $f_n: X \to [0,1]$ such that $f_n(\mathrm{Cl}(R_i)) = 0$ and $f_n(X \sim R_j) = 1$.

For any two arbitrary elements $x, y \in X$, define

$$\rho(x, y) = \sum_{1}^{\infty} 2^{-n} |f_n(x) - f_n(y)|.$$

By our definition, ρ is a nonnegative real-valued function, and we claim it is a metric. If $x = y$, then $f_n(x) = f_n(y)$ for each n and $\rho(x, y) = 0$. If $x \neq y$, by normality there is some basic element R_i containing x such that $y \notin R_i$. Using normality again, we obtain an open set V containing x such that $\mathrm{Cl}(V) \subset R_i$. But this implies there is some R_j containing x such that $\mathrm{Cl}(R_j) \subset R_i$, so $y \in X \sim R_j$. Then since R_j, $R_i = P_m$ for some m, we have $f_m(x) = 0$ and $f_m(y) = 1$, which implies $\rho(x,y) \geq 2^{-m}|f_m(x) - f_m(y)| = 2^{-m} > 0$. It is clear that $\rho(x,y) = \rho(y,x)$. If $x, y, z \in X$, then

$$\rho(x, z) = \sum_{1}^{\infty} 2^{-n} |f_n(x) - f_n(z)|$$

$$\leq \sum_{1}^{\infty} 2^{-n}(|f_n(x) - f_n(y)| + |f_n(y) - f_n(z)|)$$

$$\leq \rho(x,y) + \rho(y, z),$$

and we have shown that ρ is a metric.

It now remains to show that the topology generated by ρ is equivalent to the original topology, and we shall do this using the alternate definition of a metrizable space. Given a set M in X, let p be a limit point of M. If $\varepsilon > 0$, let N be an integer such that $2^{-N} < \varepsilon$. Because of the continuity of the f_n's, there exists a neighborhood U of p where the variation of $\sum_{1}^{N} 2^{-n}|f_n(x) - f_n(y)|$ is less than $\varepsilon/2$ for $x, y \in U$. Let $q \in U \cap M$, $q \neq p$. Then

$$\rho(p, q) = \sum_{1}^{\infty} 2^{-n} |f_n(p) - f_n(q)|$$

$$\leq \frac{\varepsilon}{2} + \sum_{N+1}^{\infty} 2^{-n} \leq \frac{\varepsilon}{2} + \frac{\varepsilon}{2} = \varepsilon,$$

and q satisfies the required conditions. Now assume that for each $\varepsilon > 0$ there is a point $q \in M$ such that $0 < \rho(p,q) < \varepsilon$, and suppose that p is not a limit point of M. Then using normality, there is a pair $P_n = R_i$, R_j such that $p \in R_i$, $M \sim p \subset X \sim R_j$. Thus we have $\rho(p, q) \geq 2^{-n}$ for all $q \in M \sim p$, which is a contradiction, so p is a limit point.

2. Every separable metric space satisfies Axioms (i)–(vi).

Let X be a separable metric space. We wish to show that X is a perfectly separable regular τ_1-space. It is clear that both $\varnothing$ (the null set) and X are open. Consider

$\{U_\alpha : \alpha \in A\}$, where each U_α is open, and let $U = \bigcup_\alpha U_\alpha$. If $x \in U$, then $x \in U_\alpha$ for some α; but since U_α is open, $x \in V_r(x) \subset U_\alpha \subset U$ for some real r, and U is open. If U_1, $U_2, \ldots, U_N$ are open sets and $U' = \bigcap_1^N U_i$, we have for each $x \in U'$ that $x \in V_{r_i}(x) \subset U_i$, $i = 1, 2, \ldots, N$. Then $x \in V_{r_0}(x) \subset U_i$ for each i if $r_0 = \min(r_1, r_2, \ldots, r_N)$. So we have $x \in V_{r_0}(x) \subset U'$, and U' is open. Hence X is at least a topological space.

If $x, y \in X$ and $\rho(x, y) = r$, then $V_{r/2}(y)$ is an open set containing y but not x, so X is a τ_1-space.

Suppose $x \in X$ and E is a closed set not containing x. Define $r = \inf\{\rho(x, y) : y \in E\}$. If $r = 0$, then x would be a limit point of E, but $x \notin E$. So $r > 0$, and the sets $U_x = V_{r/2}(x)$ and $U_E = \bigcup_{y \in E} V_{r/2}(y)$ are disjoint open sets containing x and E. Hence X is regular.

Since X is separable, it has a dense subset $A = \{a_1, a_2, \ldots\}$. Consider the collection R of all open spheres with rational radius centered at points in A. Let U be an arbitrary open set and $x \in U$, so we have $x \in V_r(x) \subset U$ for $r > 0$. Since A is dense in X, there is some a_i such that $r_0 = \rho(a_i, x) < r/2$. Because the rationals are dense in the reals, there exists a rational r_x such that $r_0 < r_x < r/2$. Then $V_{r_x}(a_i)$ is a member of R, contains x, and is contained in U, so U must be the union of elements of R. Thus R is a countable basis for X, and X is perfectly separable.

3. Show that the distance functions in Examples 2, 3, 5, and 6 satisfy the triangle inequality.

We shall need the following inequalities to complete this exercise:

(i) *Cauchy–Schwartz Inequality.*

$$\sum_1^n |a_i b_i| \le \left(\sum_1^n a_i^2\right)^{1/2} \left(\sum_1^n b_i^2\right)^{1/2} = \|a\| \, \|b\|.$$

PROOF. We have

$$0 \le \sum_1^n \left[\frac{|a_i|}{\|a\|} - \frac{|b_i|}{\|b\|}\right]^2$$

$$= \sum_1^n \frac{|a_i|^2}{\|a\|^2} + \sum_1^n \frac{|b_i|^2}{\|b\|^2} - 2\sum_1^n \frac{|a_i b_i|}{\|a\| \, \|b\|}$$

$$= 2 - 2\sum_1^n \frac{|a_i b_i|}{\|a\| \, \|b\|}$$

so that

$$\sum_1^n \frac{|a_i b_i|}{\|a\| \, \|b\|} \le 1,$$

or

$$\sum_1^n |a_i b_i| \le \|a\| \, \|b\|. \qquad \square$$

(ii) *Minkowski Inequality.* If x and y are complex numbers or real vectors $\{x_i\}$ and $\{y_i\}$, then

$$\|x + y\| \le \|x\| + \|y\|.$$

PROOF.

$$\|x + y\|^2 = \sum_1^n x_i^2 + \sum_1^n y_i^2 + 2 \sum_1^n x_i y_i$$

$$\leq \sum_1^n x_i^2 + \sum_1^n y_i^2 + 2 \sum_1^n |x_i y_i|$$

$$\leq \sum_1^n x_i^2 + \sum_1^n y_i^2 + 2 \left(\sum_1^n x_i^2 \right)^{1/2} \left(\sum_1^n y_i^2 \right)^{1/2}$$

$$= [(\sum x_i^2)^{1/2} + (\sum y_i^2)^{1/2}]^2$$

$$= [\|x\| + \|y\|]^2.$$

(iii) If $a, b, c, x, y, z,$ are nonnegative real numbers where $a \leq b + c$ and $x \leq y + z$, then

$$\sqrt{a^2 + x^2} \leq \sqrt{b^2 + y^2} + \sqrt{c^2 + z^2}. \qquad \square$$

PROOF. We first note that $bc + yz \leq \sqrt{b^2 + y^2} \sqrt{c^2 + z^2}$ by the Cauchy–Schwartz Inequality. Then

$$a^2 + x^2 \leq b^2 + c^2 + y^2 + z^2 + 2(bc + yz)$$

$$\leq (b^2 + y^2) + (c^2 + z^2) + 2(\sqrt{b^2 + y^2} \sqrt{c^2 + z^2})$$

$$= [\sqrt{b^2 + y^2} + \sqrt{c^2 + z^2}]^2.$$

We now prove the triangle inequalities.

EXAMPLE 2 (E^n). If $x, y, z \in E^n$, then

$$\rho(x,z)^2 = \sum_1^n (x_i - z_i)^2 = \sum_1^n [(x_i - y_i) + (y_i - z_i)]^2$$

$$\leq \sum_1^n (x_i - y_i)^2 + \sum_1^n (y_i - z_i)^2 + 2 \sum_1^n |(x_i - y_i)(y_i - z_i)|$$

$$\leq \sum_1^n (x_i - y_i)^2 + \sum_1^n (y_i - z_i)^2 + 2 \left(\sum_1^n (x_i - y_i)^2 \right)^{1/2} \left(\sum_1^n (y_i - z_i)^2 \right)^{1/2}$$

$$= \left[\left(\sum_1^n (x_i - y_i)^2 \right)^{1/2} + \left(\sum_1^n (y_i - z_i)^2 \right)^{1/2} \right]^2$$

$$= [\rho(x,y) + \rho(y,z)]^2.$$

Taking square roots, we have $\rho(x, z) \leq \rho(x, y) + \rho(y, z)$.

EXAMPLE 3 (Hilbert Space l_2). Let $\varepsilon > 0$ be given and x, y, z be arbitrary. Then there exists an integer N such that

$$\sum_N^\infty x_i^2 < \frac{\varepsilon^2}{4} \quad \text{and} \quad \sum_N^\infty z_i^2 < \frac{\varepsilon^2}{4}.$$

Using that $a^2 + b^2 \geq 2ab$, we have

$$\sum_N^\infty (x_i - z_i)^2 \leq 2 \sum_N^\infty x_i^2 + 2 \sum_N^\infty z_i^2 \leq \frac{2\varepsilon^2}{4} + \frac{2\varepsilon^2}{4} = \varepsilon^2.$$

We also note that $\sqrt{a + b} \leq \sqrt{a} + \sqrt{b}$. Now

$$\rho(x, z) = \sqrt{\sum_1^\infty (x_i - z_i)^2} \leq \sqrt{\sum_1^{N-1} (x_i - z_i)^2} + \sqrt{\sum_N^\infty (x_i - z_i)^2}$$

$$\leq \sqrt{\sum_1^{N-1} (x_i - z_i)^2} + \varepsilon \leq \sqrt{\sum_1^{N-1} (x_i - y_i)^2} + \sqrt{\sum_1^{N-1} (y_i - z_i)^2} + \varepsilon$$

$$\leq \sqrt{\sum_1^\infty (x_i - y_i)^2} + \sqrt{\sum_1^\infty (y_i - z_i)^2} + \varepsilon = \rho(x,y) + \rho(y,z) + \varepsilon.$$

Since ε was arbitrary, the result follows.

EXAMPLE 5 (Q_ω'). Suppose x, y, $z \in Q_\omega'$. Then using the Minkowski Inequality,

$$\rho(x, z) = \sum_1^\infty \frac{|x_i - z_i|}{2^i} \leq \sum_1^N \frac{|x_i - z_i|}{2^i} + 2^{-N}$$

$$\leq \sum_1^N \frac{|x_i - y_i|}{2^i} + \sum_1^N \frac{|y_i - z_i|}{2^i} + 2^{-N}$$

$$\leq \sum_1^\infty \frac{|x_i - y_i|}{2^i} + \sum_1^\infty \frac{|y_i - z_i|}{2^i} + 2^{-N}$$

$$= \rho(x, y) + \rho(y, z) + 2^{-N}.$$

Since 2^{-N} becomes arbitrarily small as N gets large,

$$\rho(x, z) \leq \rho(x, y) + \rho(y, z).$$

EXAMPLE 6 (Cartesian product of two metric spaces). Let $x = (x_1, x_2)$, $y = (y_1, y_2)$, and $z = (z_1, z_2)$ be points in $X \times Y$. We first note that

$$\rho_x(x_1, z_1) \leq \rho_x(x_1, y_1) + \rho_x(y_1, z_1)$$

and

$$\rho_Y(x_2, z_2) \leq \rho_Y(x_2, y_2) + \rho_Y(y_2, z_2).$$

Then by inequality (iii),

$$\rho(x,z) = \sqrt{\rho_x(x_1, z_1)^2 + \rho_y(x_2, z_2)^2}$$
$$\leq \sqrt{\rho_x(x_1, y_1)^2 + \rho_y(x_2, y_2)^2}$$
$$+ \sqrt{\rho_x(y_1, z_1)^2 + \rho_y(x_2, y_2)^2}$$
$$= \rho(x, y) + \rho(y, z).$$

SECTION VII

Diameters and Distances

Definition. For any set N, the *diameter* $\delta(N)$ is the l.u.b. (finite or infinite) of the aggregate of numbers $[\rho(x,y)]$ for x, $y \in N$.

Definition. If X and Y are nonempty sets, the *distance* $\rho(X,Y)$ is the g.l.b. of the numbers $[\rho(x,y)]$ for $x \in X$, $y \in Y$.

EXERCISES VII.

1. The distance function is continuous; i.e., if a, b are points of x, then for every $\varepsilon > 0$ there exist open sets U_a and U_b about a and b respectively such that for any $x \in U_a$ and $y \in U_b$, we have $|\rho(x,y) - \rho(a,b)| < \varepsilon$.

2. If N is compact ($N \neq \varnothing$), then there exist points x and y in N such that $\rho(x,y) = \delta(N) < \infty$.

3. If X and Y are disjoint compact nonempty sets, then there exist points $x \in X$ and $y \in Y$ such that $\rho(x,y) = \rho(X,Y) > 0$.

4. In a perfectly separable space, every set is separable. Thus every subset of a separable metric space is separable and metric.

5. Every compact metric space is separable.

6. Let A and B be subsets of a metric space X with $\text{Cl}(A) \cap B = \varnothing$ and $A \cap \text{Cl}(B) = \varnothing$. Then there exists disjoint open sets U_A and U_B containing A and B respectively.

SOLUTIONS.

1. The distance function is continuous; i.e., if a, b are points of X, then for every $\varepsilon > 0$ there exist open sets U_a and U_b about a and b respectively such that for any $x \in U_a$ and $y \in U_b$, we have $|\rho(x, y) - \rho(a,b)| < \varepsilon$.

Let a, $b \in X$ and $\varepsilon > 0$ be given. Consider the sets $U_a = V_{\varepsilon/2}(a)$ and $U_b = V_{\varepsilon/2}(b)$; i.e., if $x \in U_a$, then $\rho(x, a) < \varepsilon/2$, and if $y \in U_b$, then $\rho(y, b) < \varepsilon/2$. Then for $x \in U_a$, $y \in U_b$ we have

$$\rho(x, y) \leq \rho(x,a) + \rho(a,b) + \rho(b, y) \leq \rho(a,b) + \varepsilon$$

and

$$\rho(a,b) \leq \rho(a,x) + \rho(x, y) + \rho(y,b) \leq \rho(x, y) + \varepsilon.$$

Combining these inequalities gives

$$\rho(a,b) - \varepsilon \leq \rho(x, y) \leq \rho(a,b) + \varepsilon,$$

or

$$|\rho(x, y) - \rho(a,b)| \leq \varepsilon.$$

2. If N is compact ($N \neq \varnothing$), then there exist points x and y in N such that $\rho(x, y) = \delta(N) < \infty$.

Suppose that $\delta(N) = \infty$. If we fix $x \in N$, then for every integer n there is some $y \in N$ such that $\rho(x, y) \geq n$. But then $\{V_n(x): n = 1, 2, \ldots\}$ is an open cover for N which has no finite subcover, contradicting the compactness of N. Therefore, $\delta(N) < \infty$.

Since $\delta(N) = \text{l.u.b.}\{\rho(x, y) : x, y \in N\}$, there exist sequences $\{x_i\}$ and $\{y_i\}$ in N such that $\lim_{i \to \infty} \rho(x_i, y_i) = \delta(N)$. The set $\{x_i\}$ is either finite or infinite. If it is finite, then at least one x_i is repeated an infinite number of times; denote this x_i by x. If $\{x_i\}$ is infinite, it has a limit point in N; denote this point by x. Similarly, associated with $\{y_i\}$ is a point $y \in N$. Let $\{x_{n_i}\}$ be a subsequence of $\{x_i\}$ which converges to x, and let $\{y_{n_i}\}$ be the corresponding subsequence of $\{y_i\}$. But then $\{y_{n_i}\}$ has a subsequence $\{y_{m_i}\}$ which converges to y, and associated with it is $\{x_{m_i}\}$, which converges to x. Hence

$$\rho(x, y) = \lim_{i \to \infty} \rho(x_{m_i}, y_{m_i}) = \delta(N),$$

since ρ is continuous.

3. If X and Y are disjoint compact nonempty sets, then there exist points $x \in X$ and $y \in Y$ such that $\rho(x, y) = \rho(X, Y) > 0$.

Using the method of the preceding exercise, we obtain sequences $\{x_i\}$ and $\{y_i\}$ which converge to x and y respectively, where $x \in X$, $y \in Y$, and $\rho(x, y) = \lim_{i \to \infty} (x_i, y_i) = \rho(X,Y)$. Since $x \neq y$, $\rho(X,Y) \neq 0$.

4. In a perfectly separable space, every set is separable. Thus every subset of a separable metric space is separable and metric.

Suppose X is a perfectly separable space with basis $\{R_i\}$, and A is a subset of X. Let $\{R_{n_i}\}$ be the collection of those R_i's which intersect A. For each R_{n_i}, we pick $a_i \in R_{n_i} \cap A$ and let $A_0 = \{a_i\}$. We claim that A_0 is a countable dense set. Choose $x \in A$ such that $x \notin A_0$, and consider any open set U containing x. Then $x \in R_{n_i} \subset U$ for some i, which implies $a_i \in U$. Since U was arbitrary, x must be a limit point of A_0. Therefore A_0 is a countable dense set, and A is separable.

5. Every compact metric space is separable.

Let X be a compact metric space. Consider the collection $\{V_1(x): x \in X\}$, which is an open cover for X. There is a finite subcover of open spheres of radius 1 centered at points $x_{11}, x_{12}, \ldots, x_{1r_1}$, and we denote this set of points by A_1. Similarly, for each n, there is a finite cover for X of open spheres of radius $1/n$ centered at each point in the set $A_n = \{x_{n1}, \ldots, x_{nr_n}\}$. Let $A = \bigcup_n A_n$, and choose $x \in X \sim A$. If U is an open set containing x, then for some $\varepsilon > 0$, $x \in V_\varepsilon(x) \subset U$. Let M be an integer such that $1/M < \varepsilon$. Open spheres of radius $1/M$ centered at the points $x_{M1}, \ldots, x_{Mr_M}$ cover X, so some x_{Mi} satisfies $\rho(x, x_{Mi}) < 1/M < \varepsilon$. But then $x_{Mi} \in V_\varepsilon(x) \subset U$, so x must be a limit point of A. Hence A is a countable dense subset, and X is separable.

6. Let A and B be subsets of a metric space X with $\mathrm{Cl}(A) \cap B = \varnothing$ and $A \cap \mathrm{Cl}(B) = \varnothing$. Then there exist disjoint open sets U_A and U_B containing A and B respectively.

For $A = \varnothing$ or $B = \varnothing$ the proposition is trivially true. Whenever $A \neq \varnothing$ and $B \neq \varnothing$, then the function $f(x) = \rho(x,A) - \rho(x,B)$ is continuous for $x \in X$. Let $U_A = f^{-1}(-\infty, 0)$ and $U_B = f^{-1}(0, \infty)$.

Topological Limits

Definition. If G in an infinite collection of sets in X, the set I of all points $x \in X$ such that every open set about x meets all but a finite number of elements of G is called the *limit inferior* of G (i.e., $\underline{\text{Lim}}\ G = I$). The set L of all $x \in X$ such that every open set about x meets infinitely many elements of G is called the *limit superior* of G, (i.e., $\overline{\text{Lim}}\ G = L$).

In case $I = L$ for a given collection or sequence G, then G is said to *converge* to this common set. ($I = L = \varnothing$ is possible.)

EXAMPLE. In the Euclidean plane let $G_i = \{(x,1/i) | 0 \le x \le 2\}$ for i an odd positive integer, and let $G_i = \{(x,1/i) | 1 \le x \le 3\}$ for i an even positive integer. Then $L = \{(x,0) | 0 \le x \le 3\}$ and $I = \{(x,0) | 1 \le x \le 2\}$.

EXERCISES VIII.

1. If G^* is any infinite subcollection of an infinite collection G, we have

$$\underline{\lim}\ G \subset \underline{\lim}\ G^* \subset \overline{\lim}\ G^* \subset \overline{\lim}\ G.$$

Thus, if G converges, then so does any subcollection.

2. In a perfectly separable space, every infinite collection of sets contains a convergent infinite subcollection.

3. If $\{A_i\}$ is a sequence of sets in a τ_1-space with $\overline{\lim}\ A_i = L$ and with $\text{Cl}(\bigcup_1^\infty A_i)$ compact, then if U is any open set containing L, there exists an integer N such that $A_n \subset U$ for $n > N$.

SOLUTIONS.

1. If G^* is any infinite subcollection of an infinite collection G, we have

$$\underline{\lim} \, G \subset \underline{\lim} \, G^* \subset \overline{\lim} \, G^* \subset \overline{\lim} \, G.$$

Thus, if G converges, then so does any subcollection.

Using the fact that both G and G^* are infinite collections, the above inclusions follow immediately from the definitions. If G converges, then $\underline{\lim} \, G = \overline{\lim} \, G = \underline{\lim} \, G^* = \lim G^*$, and G^* converges to the same limit as G.

2. In a perfectly separable space, every infinite collection of sets contains a convergent infinite subcollection.

Suppose $G = [G_\alpha]$ is an infinite collection of subsets of the perfectly separable space X and let $[R_i]$ be a countable basis for X. Let $[G_{\alpha'}]$ be an infinite subcollection of $[G_\alpha]$ such that $R_1 \cap \overline{\lim} \, G_{\alpha'} = \varnothing$ if such a subcollection exists. Otherwise, let $[G_\alpha^1] = [G_\alpha]$. Similarly, for each n, let $[G_\alpha^n]$ be an infinite subcollection of $[G_\alpha^{n-1}]$ satisfying $R_n \cap \overline{\lim} \, G_\alpha^n = \varnothing$ if such exists. If not, let $[G_\alpha^n] = [G_\alpha^{n-1}]$. Since each $[G_\alpha^n]$ is infinite, we can choose sets $A_n \in [G_\alpha^n]$ so that $A_n \neq A_m$ if $n \neq m$, thus guaranteeing that $[A_n]$ is an infinite subcollection of $[G_\alpha]$.

We shall now show that $[A_n]$ converges. Suppose not; i.e., suppose that there is some $x \in \overline{\lim} \, A_n \sim \underline{\lim} \, A_n$. Then there exists an open set U containing x whose complement contains an infinite number of A_n's. Since $x \in R_j \subset U$ for some j, R_j^c also contains an infinite number of A_n's. Let $\{A_{n_k}\}$ be those A_n's missed by R_j, where $n_k \geq j$ for all k. Then $R_j \cap \overline{\lim} \, A_{n_k} = \varnothing$. We also note that $\{A_{n_k}\}$ is infinite and $\{A_{n_k}\} \subset [G_\alpha^{j-1}]$. Going back to the construction of $[G_\alpha^j]$, we see that $[G_\alpha^{j-1}]$ does in fact have an infinite subcollection $[G_\alpha^j]$ such that $R_j \cap \overline{\lim} \, G_\alpha^j = \varnothing$. But for $n \geq j$, $A_n \subset [G_\alpha^j]$, implying that $\overline{\lim} \, A_n \subset \overline{\lim} \, G_\alpha^j$ and hence $R_j \cap \overline{\lim} \, A_n = \varnothing$. But $x \in R_j \cap \overline{\lim} \, A_n \neq \varnothing$; this contradiction establishes the proposition.

3. If $\{A_i\}$ is a sequence of sets in a τ_1-space with $\overline{\lim} \, A_i = L$ and $\mathrm{Cl}(\bigcup_1^\infty A_i)$ compact, then if U is any open set containing L, there exists an integer N such that $A_n \subset U$ for $n > N$.

Assume that the proposition is not true. Then we can find an open set U containing L and a sequence of points $p_1, p_2, \ldots$ such that $p_i \in A_{n_i}$, where $n_{i+1} > n_i$, and $p_i \notin U$. Since $\mathrm{Cl}(\bigcup_1^\infty A_i)$ is compact, there exists a point p which is either a limit point of $\{p_i\}$ or equal to some p_i which appears an infinite number of times. In either case, $p \in L$. Since U is then an open set containing p, U must contain at least one p_i, which leads to a contradiction, since by construction, no p_i is contained in U. Therefore, the proposition must be true.

Relativization

Definition. If M is any subset of a topological space X, then the *relative topology* in M induced by the topology on X is that given by defining a set M' to be open in M if and only if there exists an open set U in X such that $M' = M \cap U$.

Such a subset M' of M is said to be *open in M* or open relative to M.

Definition. A set $N \subset M$ is by definition *closed in M* if and only if it contains all of its limit points that belong to M. An equivalent definition is that there exists a closed set C in X such that $N = M \cap C$. (Take $C = \mathrm{Cl}(N)$.)

SECTION X

Connected Sets

Definition. For any set M, a representation $M = M_1 \cup M_2$ of M as the union of two nonempty disjoint sets, neither one containing a limit point of the other, is called a *separation*. The sets M_1 and M_2 in a separation are called *separated sets*.

Definition. A set M is said to be *connected* if and only if it admits no separation whatever.

Definition. Two nonempty sets A and B are separated if and only if $A \cap \mathrm{Cl}(B) = \mathrm{Cl}(A) \cap B = \varnothing$.

Definition. If M is any set, a *component* of M is a nonempty connected subset of M which is not contained in any other connected subset of M, i.e., a maximal connected subset of M.

Definition. For $\varepsilon > 0$, an *ε-chain* is a finite sequence of points $x_1, x_2, \ldots, x_n$ where $\rho(x_i, x_{i+1}) < \varepsilon$, $i = 1, 2, \ldots, n - 1$. A set M is *well chained* if and only if for every $\varepsilon > 0$, each pair of points in M can be joined by an ε-chain of points of M.

EXERCISES X.

1. If $[G]$ is any collection of connected sets with $\bigcap_{G \in [G]} G \neq \varnothing$, then $\bigcup_{G \in [G]} G$ is connected.

2. If M is a connected set, so also is any set M_0 such that $M \subset M_0 \subset \mathrm{Cl}(M)$. In particular, $\mathrm{Cl}(M)$ is connected.

3. A set M is connected if and only if no proper (nonempty) subset of M is both open and closed in M.

4. If N is a connected subset of a connected set M and $M \sim N = M_1 \cup M_2$, M_1 and M_2 being separated sets, then $M_1 \cup N$ and $M_2 \cup N$ are connected.

5. Any component of a set K is closed relative to K.

6. If A is any subset of a set M in a metric space, then for every $\varepsilon > 0$ the set M_A of all points of M which can be joined to a point $a \in A$ by an ε-chain of points of M is both open and closed in M.

7. Every connected set is well chained.

8. Every compact, well-chained set is connected.

SOLUTIONS.

1. If $[G]$ is any collection of connected sets with $\bigcap_{G \in [G]} G \neq \varnothing$, then $\bigcup_{G \in [G]} G$ is connected.

Suppose $\bigcup G$ were not connected. Then $\bigcup G = A_1 \cup A_2$, where $\mathrm{Cl}(A_1) \cap A_2 = A_1 \cap \mathrm{Cl}(A_2) = \varnothing$, $A_1 \neq \varnothing \neq A_2$. Since each $G \in [G]$ is connected, $G \subset A_1$ or $G \subset A_2$; otherwise, $A_1 \cap G$ and $A_2 \cap G$ would separate G. Let $[G^i] = \{G \in [G] : G \subset A_i\}$ for $i = 1, 2$. Then

$$\bigcap_{G \in [G]} G = \left(\bigcap_{G \in [G^1]} G \right) \cap \left(\bigcap_{G \in [G^2]} G \right)$$

$$\subset A_1 \cap A_2 = \varnothing,$$

which contradicts the assumption that $\bigcap_{G \in [G]} G \neq \varnothing$. Thus $\bigcup G$ is connected.

2. If M is a connected set, so also is any set M_0 such that $M \subset M_0 \subset \mathrm{Cl}(M)$. In particular, $\mathrm{Cl}(M)$ is connected.

Assume that M_0 is not connected and $M_0 = A \cup B$ is a separation. Then M is contained in A or B, since otherwise $M = (M \cap A) \cup (M \cap B)$ would be a separation of M. Suppose that $M \subset A$, and then let $x \in B$. Because $B \subset \mathrm{Cl}(M)$, and $\mathrm{Cl}(M) \subset \mathrm{Cl}(A)$, we have $x \in B \cap \mathrm{Cl}(A) \neq \varnothing$ and $M_0 = A \cup B$ was not a separation. Hence M_0 must be connected.

3. A set M is connected if and only if no proper (nonempty) subset of M is both open and closed in M.

If M is a set with a proper (nonempty) subset N which is both open and closed relative to M, then $M = N \cup (M \sim N)$ is a separation for M. Hence M is not connected.

Suppose that M is a set having no proper nonempty subset which is both open and closed in M, and assume that M is not connected. Then $M = A \cup B$, where $\mathrm{Cl}(A) \cap B = A \cap \mathrm{Cl}(B) = \varnothing$. Consider one of these sets, say A. If $x \in M$ and x is a limit point of A, then $x \notin B$, so that $x \in A$, and A is closed in M. Let $U = X \sim \mathrm{Cl}(B)$. Then U is open in X, and $U \cap M = A$, so we have A being both open and closed in M. This contradicts our assumption, and M must be connected.

4. If N is a connected subset of a connected set M and $M \sim N = M_1 \cup M_2$, M_1 and M_2 being separated sets, then $M_1 \cup N$ and $M_2 \cup N$ are connected.

Suppose that $M_1 \cup N$ is not connected and $M_1 \cup N = A \cup B$, where A and B are separated sets. Since N is connected, $N \subset A$ or $N \subset B$; otherwise $N \cap A$ and $N \cap B$ would separate N. Assume $N \subset A$, so that $B \subset M_1$. But then $M = (M_2 \cup A) \cup B$ is a separation for M, contradicting the assumption that M is connected. Hence $M_1 \cup N$ must be connected and, by symmetry, so also is $M_2 \cup N$.

5. Any component of a set K is closed relative to K.

Consider a component M of K, and let M_0 be the closure of M in K; i.e., $M_0 = \mathrm{Cl}(M) \cap K$. Then $M \subset M_0 \subset \mathrm{Cl}(M)$, and M_0 is connected, so that $M = M_0$ and M is closed in K.

6. If A is any subset of a set M in a metric space, then for every $\varepsilon > 0$ the set M_A of all points of M which can be joined to a point $a \in A$ by an ε-chain of points of M is both open and closed in M.

(Note: it is understood that in problems concerning ε-chains we are in a metric space.) Let $\varepsilon > 0$ be given, and let M_A be as defined above. If $x \in M$ is a limit point of

M_A, then there is some element $y \in M_A$ satisfying $\rho(x, y) < \varepsilon$. Let $y, y_1, \ldots, y_N$, where $y_N \in A$, be an ε-chain joining y to A. Then $x, y, y_1, \ldots, y_N$ is an ε-chain joining x to A, and $x \in M_A$. Hence M_A is closed in M.

Now let z be a limit point of M_A^c. If $z \in M_A$, then there exists an ε-chain $z, z_1, \ldots, z_M$ joining z to A. Also, for the given ε, there is a point $w \in M_A^c$ such that $\rho(z, w) < \varepsilon$. But then w is joinable to A by an ε-chain and $w \in M_A \cap M_A^c$, which is clearly a contradiction. Hence $z \in M_A^c$, M_A^c is closed in M, and M_A is open in M.

7. Every connected set is well chained.

Let M be a connected set, and suppose M is not well chained. Then there exist two points $x, y \in M$ which cannot be joined by an ε-chain of points in M for some $\varepsilon > 0$. Consider M_x, the set of all points which can be joined to x by an ε-chain of points in M. Then M_x is a proper subset of M which is both open and closed in M, implying that M is not connected. Therefore, M must be well chained.

8. Every compact, well-chained set is connected.

Consider a compact, well-chained set M, and suppose it isn't connected. Then $M = A \cup B$, where A and B are both open and closed in M. Suppose $[U_\alpha]$ is an open cover for A, and $B = M \cap V$, where V is open in X. Then $[U_\alpha] \cup V$ is an open cover for M, which must have a finite subcover $U_1, U_2, \ldots, U_N, V$, and $U_1, U_2, \ldots, U_N$ must cover A. Therefore A is compact and, by symmetry, so also is B. Let $\rho(A, B) = d > 0$. Pick points $a \in A$ and $b \in B$. Then a and b cannot be joined by an ε-chain for $\varepsilon < d$, contradicting the assumption that M is well chained. Hence M is connected.

SECTION XI

Connectedness of Limit Sets and Separations

EXERCISES XI.

1. If $\{A_i\}$ is a sequence of connected sets in a Hausdorff space X satisfying:

   (a) $\text{Cl}(\bigcup_1^\infty A_i)$ is compact,

   (b) $\varliminf A_i \neq \varnothing$,

   then $\varlimsup A_i$ is connected.

2. If A and B are nonempty, closed, disjoint subsets of a compact set K in a Hausdorff space such that no component of K meets both A and B, then there exists a separation $K = K_a \cup K_b$ with $A \subset K_a$ and $B \subset K_b$.

3. In a Hausdorff space, if $M_1 \supset M_2 \supset \cdots$ is a monotone decreasing sequence of nonempty, compact, connected sets, then the intersection of all of the sets is non-empty, compact, and connected.

4. Given a sequence $\{A_i\}$ in a metric space such that

   (a) $\text{Cl}(\bigcup_1^\infty A_i)$ is compact

   (b) $\varliminf A_i \neq \varnothing$,

   (c) for each i, any pair of points of A_i lies in an ε-chain in A_i and $\varepsilon_i \to 0$ as $1/i$, then $\varlimsup A_i$ is connected.

5. In a metric space, if two points a and b of a compact set K can be joined in K by an ε-chain for every $\varepsilon > 0$, then a and b lie together in the same component of K.

SOLUTIONS.

1. If $\{A_i\}$ is a sequence of connected sets in a Hausdorff space X satisfying

(a) $Cl(\bigcup_1^\infty A_i)$ is compact,
(b) $\underline{\lim}\, A_i \neq \varnothing$,

then $\overline{\lim}\, A_i$ is connected.

Consider the relative topology in $Cl(\bigcup_1^\infty A_i)$. Since X is Hausdorff and $Cl(\bigcup_1^\infty A_i)$ is compact, the relative topology is normal. We first note that $\overline{\lim}\, A_i = L$ is closed in X. This follows from the fact that if x is a limit point of L, then every open set containing x must contain a point of L and hence must intersect an infinite number of the A_i's, so that $x \in L$. Also, $L \subset Cl(\bigcup_1^\infty A_i)$, and therefore L is a closed set in $Cl(\bigcup_1^\infty A_i)$. Now suppose L is not connected, and $L = A \cup B$ is a separation for L. Since L is closed in $Cl(\bigcup_1^\infty A_i)$, A and B are closed disjoint sets in $Cl(\bigcup_1^\infty A_i)$. By the normality of the relative topology, there exist disjoint sets U_A and U_B, open relative to $Cl(\bigcup_1^\infty A_i)$, which contain A and B respectively. Then $U_A \cup U_B$ is an open set in $Cl(\bigcup_1^\infty A_i)$ which contains L, so that $A_n \subset U_A \cup U_B$ for $n > N$. We note here that both U_A and U_B intersect an infinite number of A_i's, since each contains points of L. Suppose that, for each $n\,(n > N)$, $A_n \subset U_A$ or $A_n \subset U_B$. Then if $x \in L$, we have $x \in U_A$ or $x \in U_B$: in either case we have found an open set containing x which doesn't intersect all but a finite number of A_i's. But then $I = \lim A_i = \varnothing$. Therefore there exists at least one $n > N$, say $n = m$, such that $A_m = (A_m \cap U_A) \cup (A_m \cap U_B)$, where $A_m \cap U_A \neq \varnothing \neq A_m \cap U_B$. But then we have that A_m is not connected, which is a contradiction. Thus L must be connected.

2. If A and B are nonempty, closed, disjoint subsets of a compact set K in a Hausdorff space such that no component of K meets both A and B, then there exists a separation $K = K_a \cup K_b$ with $A \subset K_a$ and $B \subset K_b$.

Let $a \in A$ and $b \in B$. The points a and b do not lie in the same component; hence K is not connected. Consider all possible separations $K = A_\alpha \cup B_\alpha$ where $a \in A_\alpha$ (b is not necessarily contained in B_α). We claim that $\bigcap A_\alpha$ is connected. Suppose not; i.e., suppose that $\bigcap A_\alpha = M \cup N$ is a separation where $a \in M$. By the normality of the relative topology in K, we can find disjoint open sets U_M and U_N in K which contain M and N respectively. The set $H = K \sim (U_M \cup U_N)$ is closed and compact and covered by the collection $[B_\alpha]$, so there exists some finite subcollection $B_{\alpha_1}, B_{\alpha_2}, \ldots, B_{\alpha_n}$ with $H \subset \bigcup_1^n B_{\alpha_i}$. Each B_{α_i} is open and closed in K, so that $\bigcup_1^n B_{\alpha_i}$ is open and closed in K. Furthermore $L = \bigcup_1^n B_{\alpha_i} \bigcup U_N$ is clearly open in K, and it is also closed in K, since if $x \in Cl(U_N) - U_N$, then x is in some B_{α_i}. Thus $K = L \cup (K \sim L)$ is a separation of K, where $a \in (K \sim L)$ and $N \subset L$, which is impossible, since a and N were contained in $\bigcap A_\alpha$ and hence never separated in K. Therefore $\bigcap A_\alpha$ is connected.

Returning to the points a and b, we see that since $a \in \bigcap A_\alpha$, b cannot be contained in $\bigcap A_\alpha$; otherwise a and b would lie in the same component. Then for some α, $K = A_\alpha \cup B_\alpha$, where $a \in A_\alpha$ and $b \in B_\alpha$.

Now fix $b \in B$. For each $a \in A$ we have the separation $K = A_\alpha \cup B_\alpha$, $a \in A_\alpha$, $b \in B_\alpha$. Then $\{A_\alpha\}$, $a \in A$, is an open cover for A, so that $A \subset \bigcup_1^n A_{\alpha_i} = U_b$, which is both open and closed in K and does not contain b. Hence we have a separation $K = U_b \cup V_b$ where $A \subset U_b$ and $b \in V_b$. Now cover B with $\{V_b, b \in B\}$. We have $B \subset \bigcup_1^n V_{b_i} = K_b$, K_b being an open and closed set in K which contains B but does not intersect A. Hence $K = K_a \cup K_b$ is the required separation, where $K_a = K \sim K_b$.

3. In a Hausdorff space, if $M_1 \supset M_2 \supset \cdots$ is a monotone decreasing sequence of nonempty, compact, connected sets, then the intersection of all of the sets is nonempty, compact, and connected.

We have previously shown that $\bigcap M_i \neq \varnothing$ and is closed; it follows immediately that $\bigcap M_i$ is compact. Therefore we need only show that $\bigcap M_i$ is connected. We establish this if we note that $\underline{\lim}\, M_i = \overline{\lim}\, M_i = \bigcap M_i$ and $\bigcup M_i = M_1$ is compact.

4. Given a sequence $\{A_i\}$ in a metric space such that

(a) $\mathrm{Cl}(\bigcup_1^\infty A_i)$ is compact,
(b) $\underline{\lim}\, A_i \neq \varnothing$,
(c) for each i, any pair of points of A_i lies in an ε_i-chain in A_i and $\varepsilon_i \to 0$ as $1/i$,

then $\overline{\lim}\, A_i$ is connected.

Suppose that $L = \overline{\lim}\, A_i$ is not connected. Then let $L = A \bigcup B$ be a separation for L. Since L is a closed subset of $\mathrm{Cl}(\bigcup_1^\infty A_i)$ and both A and B are closed in L, we know that A and B are disjoint compact sets. Hence $\rho(A, B) = 3d > 0$, so that $\rho(V_d(A), V_d(B)) > d$. Let N be an integer such that for $n > N$, $A_n \subset V_d(L) = V_d(A) \bigcup V_d(B)$. Since $I \subset L$ $(I = \underline{\lim}\, A_i)$, I must intersect either A or B, say A. Then for all $n > N$, $A_n \cap V_d(A) \neq \varnothing$, because $V_d(A)$ is an open set about points in I. Also we note that for for an infinite number of integers $n > N$, $A_n \cap V_d(B) \neq \varnothing$, since $B \subset L$. There exists some integer K such that $\varepsilon_K < d$ and $A_K \cap V_d(A) \neq \varnothing \neq A_K \cap V_d(B)$. Choose points $a, b \in A_K$ where $a \in V_d(A)$ and $b \in V_d(B)$. Then a and b cannot be joined by an ε_K-chain, since $\rho(V_d(A), V_d(B)) > d > \varepsilon_K$. But this contradicts (c), so that $L = \overline{\lim}\, A_i$ must be connected.

5. In a metric space, if two points a and b of a compact set K can be joined in K by an ε-chain for every $\varepsilon > 0$, then a and b lie together in the same component of K.

Suppose a and b do not lie in the same component of K. Then $\{a\}$ and $\{b\}$ are nonempty closed disjoint subsets of K such that no component of K meets both $\{a\}$ and $\{b\}$, and there exists a separation $K = K_a \cup K_b$ where $a \in K_a$ and $b \in K_b$. Because K_a and K_b are disjoint compact sets, $\rho(K_a, K_b) = d > 0$. But then a and b cannot be joined by an ε-chain for $0 < \varepsilon < d$. Hence a and b must lie in the same component of K.

Continua

Definition. A compact connected set is called a *continuum*.

Definition. If G is any open set, the set $\mathrm{Cl}(G) \sim G$ is called the boundary of G (or frontier of G) and is denoted $\mathrm{Fr}(G)$ or ∂G. More generally, for an arbitrary set A, $\mathrm{Fr}(A) = \mathrm{Cl}(A) \cap \mathrm{Cl}(A^c)$.

Note. The boundary of any open set is a closed set, since $\mathrm{Fr}(G) = G^c \cap \mathrm{Cl}(G)$.

EXERCISES XII.

1. If N is any continuum in a Hausdorff space and G is any open set such that $N \cap G \neq \varnothing \neq N \cap G^c$, then every component of $N \cap \mathrm{Cl}(G)$ meets $\mathrm{Fr}(G)$.

2. In a Hausdorff space X, if N and G are sets satisfying the above conditions, then every component of $N \cap G$ has a limit point on $\mathrm{Fr}(G)$.

3. In a Hausdorff space, no continuum is the union of a countable number (greater than 1) of closed disjoint sets.

SOLUTIONS.

1. If N is any continuum in a Hausdorff space and G is any open set such that $N \cap G \neq \varnothing \neq N \cap G^c$, then every component of $N \cap \text{Cl}(G)$ meets $\text{Fr}(G)$.

Suppose that some component A of $N \cap \text{Cl}(G)$ does not intersect $\text{Fr}(G)$. If we let $B = \text{Fr}(G) \cap N$, then $B \neq \varnothing$; otherwise $N = (N \cap G) \cup (N \cap G^c)$ would be a separation for N. Since $B \subset N \cap \text{Cl}(G)$, we have that A and B are nonempty, closed, disjoint subsets of a compact set $N \cap \text{Cl}(G)$ where no component of $N \cap \text{Cl}(G)$ intersects both A and B. Then there exists a separation $N \cap \text{Cl}(G) = K_a \cup K_b$ where $A \subset K_a$ and $B \subset K_b$. Consider the representation $N = ((N \sim \text{Cl}(G) \cup K_b) \cup K_a$. If p is a limit point of $N \sim \text{Cl}(G)$, then $p \notin K_a$, for suppose that p were contained in $K_a \subset N \cap \text{Cl}(G)$. We know that $p \in G$, so that $p \in \text{Fr}(G) \cap N = B$, but then $p \in K_b$ which is impossible since $K_a \cap K_b = \varnothing$. Hence $p \notin K_a$. Thus the representation $N = ((N \sim \text{Cl}(G) \cup K_b) \cup K_a$ is a separation, which contradicts the connectedness of N; so that every component of $N \cap \text{Cl}(G)$ meets $\text{Fr}(G)$.

2. In a Hausdorff space X, if N and G are sets satisfying the above conditions, then every component of $N \cap G$ has a limit point on $\text{Fr}(G)$.

Suppose the proposition is false, and A is a component of $N \cap G$ having no limit point in $\text{Fr}(G)$. Then, since A is closed relative to $N \cap \text{Cl}(G)$, A must be compact in X. Hence there exists an open set H such that $A \subset H \subset \text{Cl}(H) \subset G$. Let C be the component of $N \cap \text{Cl}(H)$ which contains A. Then C intersects $\text{Fr}(H)$; i.e., C contains a point of H^c. Hence $C \neq A$, and A is not a maximal connected set in $N \cap G$, since $A \subset C \subset N \cap H \subset N \cap G$. This contradiction establishes that the proposition is true.

3. In a Hausdorff space, no continuum is the union of a countable number (greater than 1) of closed disjoint sets.

Suppose that a continuum $M = \bigcup K_i$ where the K_i's are closed, disjoint sets. Since the relative topology in M is normal, we can find a set G_1, open relative to M, which contains K_2 and such that $\text{Cl}(G_1) \cap K_1 = \varnothing$. Let M_1 be a component of $\text{Cl}(G_1)$ which intersects K_2. (Note that M_1 is itself a continuum.) Then $M_1 \cap \text{Fr}(G_1) \neq \varnothing$; i.e., M_1 contains a point $p \in \text{Fr}(G_1)$ such that $p \notin G_1$ and $p \notin K_1$. Hence M_1 intersects some K_i for $i > 2$. Let K_{n_2} be the first K_i for $i > 2$ which intersects M_1, and let G_2 be an open set relative to M satisfying $K_{n_2} \subset G_2$ and $\text{Cl}(G_2) \cap K_2 = \varnothing$. Then let M_2 be a component of $M_1 \cap \text{Cl}(G_2)$ which contains a point of K_{n_2}. Again, we have $M_2 \cap \text{Fr}(G_2) \neq \varnothing$, and M_2 contains some point $p \in \text{Fr}(G_2)$ such that $p \notin G_2$, $p \notin K_1 \cup K_2$. Hence M_2 intersects some K_i for $i > n_2$, and $M_2 \cap K_i = \varnothing$ for $i < n_2$. Let K_{n_3} be the first K_i for $i > n_2$ which intersects M_2; then by methods similar to the above we can find a continuum M_3 such that $M_3 \subset M_2 \subset M_1$ and M_3 intersects some K_i with $i > n_3$, but $M_3 \cap K_i = \varnothing$ for $i < n_3$.

In this manner we obtain a sequence of subcontinua of M: $M_1 \supset M_2 \supset M_3 \cdots$, such that for each j, $M_j \cap K_i = \varnothing$ for $i < n_j$ and $n_j \to \infty$ as $j \to \infty$. We know that $\bigcap M_i \neq \varnothing$. Also $(\bigcap M_i) \cap K_j = \varnothing$ for all j, so that $(\bigcap M_i) \cap (\bigcup K_j) = \varnothing$ or $(\bigcap M_i) \cap M = \varnothing$. But $\bigcap M_i \subset M$, which contradicts the fact that $\bigcap M_i$ is nonempty. Hence no continuum such as M can exist.

Irreducible Continua and a Reduction Theorem

Definition. A set K is said to be *irreducible* with respect to a property P if and only if K has property P but no nonempty proper closed subset of K has property P.

Definition. A set M which is irreducible with respect to the property of being a continuum containing two points a and b (or a closed set K) is called an *irreducible continuum* from a to b.

Definition. A property P is said to be *inductive* provided that when each set in a monotone decreasing sequence $A_1 \supset A_2 \supset \cdots$ of compact sets has property P, so also does $\bigcap_1^\infty A_n$.

Note. When considering inductive properties, we will assume that all space are Hausdorff and hence $\bigcap_1^\infty A_n \neq \varnothing$.

EXAMPLES. (inductive properties).

(1) Being nonempty.
(2) Being connected.

EXERCISES XIII.

1. *Brouwer Reduction Theorem*: If P is an inductive property in a perfectly separable Hausdorff space, then any compact set K (nonempty) having property P has a nonempty closed subset which has property P irreducibly.

2. In a perfectly separable Hausdorff space, if K is any nonempty closed subset of a continuum M, then M contains a subcontinuum which is irreducible about K.

SOLUTIONS.

1. *Brouwer Reduction Theorem*: In a perfectly separable Hausdorff space, if P is an inductive property, then any compact set K having property P has a nonempty closed subset which has property P irreducibly.

Let $\{R_i\}$ be a countable basis for X, and assume that K is not itself irreducible with respect to P. Let n_1 be the first integer such that $K \sim R_{n_1}$ has a nonempty proper closed subset A_1 which has property P. The integer n_1 must exist, since K does contain some nonempty proper subset H which has property P. Thus $K \sim H$ is open in K and nonempty; i.e., $K \sim H = K \cap U \neq \varnothing$, where U is open. If $x \in K \cap U$, then $x \in R_m \subset U$ and $R_m \cap H \subset U \cap H = \varnothing$, so that n_1 exists and is less than or equal to m. Now let n_2 be the first integer larger than n_1 such that $A_1 \sim R_{n_2}$ has a nonempty proper closed subset A_2 with property P. If such an integer does not exist, then by the same argument used for K, A_1 must be irreducible relative to P. In general, for every integer k, let n_k be the smallest integer larger than n_{k-1} such that $A_{k-1} \sim R_{n_k}$ contains a nonempty proper closed subset A_k with property P. If for some integer m, n_m does not exist, then A_{m-1} is irreducible relative to property P.

Suppose that n_k exists for all k, since otherwise the proposition is trivially true. Then we have a monotone decreasing sequence $A_1 \supset A_2 \supset \cdots$ of nonempty compact sets with property P. If $A = \bigcap A_i$ then since X is Hausdorff and P is inductive, A is a nonempty compact (closed) subset of K with property P. Moreover, A is irreducible relative to P, for suppose that B is a proper closed subset of A which has property P. Then, since $A \sim B$ is a nonempty set which is open relative to A, there is some R_k such that $R_k \cap A \neq \varnothing$ while $R_k \cap B = \varnothing$. Since $R_{n_i} \cap A = \varnothing$ for all i, we have $k \neq n_i$ for any i, so that there exists an integer j such that $n_j < k < n_{j+1}$. Since $B \subset A$, B is certainly contained in A_j. Returning to the definition of n_{j+1}, we see that n_{j+1} is the least integer larger than n_j such that $A_j \sim R_{n_{j+1}}$ contains a closed nonempty subset with property P. But this is a contradiction, since $A_j \sim R_k$ contains B, so that n_{j+1} would have to be less than or equal to k, and we have $n_{j+1} > k$. Hence the subset B cannot exist, and A is irreducible with respect to P.

2. In a perfectly separable Hausdorff space, if K is any nonempty closed subset of a continuum M, then M contains a subcontinuum which is irreducible about K.

This property follows immediately from the Brouwer Reduction Theorem if we note that the property of being a continuum containing K is inductive.

Locally Connected Sets

Definition. A set M is *locally connected at a point* $p \in M$ provided that if U_p is any open set about p, then there exists an open set V_p, $p \in V_p \subset U_p$, such that each point of $M \cap V_p$ lies together with p in a connected subset of $M \cap U_p$.

Equivalently, $M \cap V_p$ lies in the component of $M \cap U_p$ containing p.

EXAMPLES. Consider the following subsets of the Euclidean plane R^2 with the relative topology:

(i) $I = \{(x,0)|0 \le x \le 1\}$.
(ii) $I_2 = \{(x,y)|0 \le x \le 1, 0 \le y \le 1\}$.
(iii) Using polar coordinates, for n a positive integer, let $I_n = \{(x,\pi/2n)|0 \le x \le 1\}$ and $I_0 = \{(x,0)|0 \le x \le 1\}$. The set $H = \bigcup_{n=0}^{\infty} I_n$ with the relative topology fails to be locally connected at all points of I_0 except $(0,0)$.

Definition. A set M is *locally connected* if and only if it is locally connected at each of its points.

Definition. A connected relatively open subset of a set M is called a *region* in M.

Definition. A set L is *locally compact* if and only if each point $p \in L$ lies in an open set U_p such that $L \cap \mathrm{Cl}(U_p)$ is compact.

EXAMPLE. A line in R^2 is locally compact even though not compact.

Definition. A locally compact, connected Hausdorff space is called a *generalized continuum*.

EXERCISES XIV.

1. A set M is locally connected if and only if each component of an open subset of M (open relative to M) is itself open in M.

2. A set M is locally connected if and only if for each point p in M and each open set U_p about p, there exists a region R_p in M containing p and lying in U_p. Equivalently in a metric space, each point of M lies in an arbitrarily small region in M.

3. Every region in a locally connected generalized continuum is itself a locally connected generalized continuum.

SOLUTIONS.

1. A set M is locally connected if and only if each component of an open subset of M (open relative to M) is itself open in M.

Let M be a locally connected set, U an open subset of M, and K a component of U. Choose a point $p \in K$. Since U is open in M, $U = M \cap U_p$, where U_p is an open set containing p. By the local connectedness of M, there is an open set V_p such that $p \in V_p \subset U_p$ and $M \cap V_p$ lies in a connected subset of $M \cap U_p$. But then $M \cap V_p$ must be contained in K; otherwise K is not maximal. Hence we have found that $M \cap V_p$ is an open subset of M satisfying $p \in M \cap V_p \subset K$, which is sufficient to show that K is open in M.

Now suppose that each component of an open subset of M is itself open in M. Let $p \in M$ and U_p be an open set containing p. If K is the component of $M \cap U_p$ containing p, then $K = M \cap V$, where V is open. The open set $V_p = V \cap U_p$ has the properties that $p \in V_p \subset U_p$ and $V_p \cap M$ lies in a connected subset of $U_p \cap M$, since $V_p \cap M \subset K$. Hence M is locally connected at the point p, and so is locally connected.

2. A set M is locally connected if and only if for each point p in M and each open set U_p about p, there exists a region R_p in M containing p and lying in U_p. Equivalently in a metric space, each point of M lies in an arbitrarily small region in M.

Assume that M is locally connected, and let $p \in M$, U_p an open set containing p. If K is the component in $M \cap U_p$ which contains p, then K is the required region in M.

Suppose that for each $p \in M$ and any open set U_p containing p, we can find a region R_p satisfying the above conditions. Since R_p is open in M, $R_p = V \cap M$, where V is open. Then, if $V_p = V \cap U_p$, we have $p \in V_p \subset U_p$ and $V_p \cap M \subset R_p$, so that $V_p \cap M$ lies in a connected subset of U_p. Hence M is locally connected at the point p, and so is locally connected.

3. Every region in a locally connected generalized continuum is itself a locally connected generalized continuum.

If M is a locally connected generalized continuum, we will consider only the relative topology in M. Let R be a region in M. We wish to show that R is locally connected and locally compact. Suppose $p \in R$ and U_p is an open set containing p. Since M is locally connected, and $U_p \cap R$ is open, there exists an open set V_p such that $p \in V_p \subset U_p \cap R \subset U_p$ and V_p lies in a connected subset of $U_p \cap R$. But then $V_p \subset R$, so that $V_p \cap R = V_p$ also lies in a connected subset of $U_p \cap R$ and R is locally connected.

If $p \in R$, then since M is locally compact, there is an open set U such that $p \in U$ and $\mathrm{Cl}(U)$ is compact. Since X is Hausdorff and $\mathrm{Fr}(U)$ is compact, there is an open set V in M with $p \in V \subset \mathrm{Cl}(V) \subset U \cap R$. The open set $R \cap V$ contains p, and $\mathrm{Cl}(R \cap V)$ is compact and in R.

SECTION XV
Property S and Uniformly Locally Connected Sets

Definition. A set M in a metric space is said to have *property S* if and only if for every $\epsilon > 0$, M is the union of a finite number of connected sets, each of diameter less than ϵ.

Definition. A set M, also in a metric space, is said to be *uniformly locally connected* if and only if for every $\epsilon > 0$ there exists a $\delta > 0$ such that any pair of points x, y of M with $\rho(x, y) < \delta$ lie together in a connected subset of M of diameter less than ϵ.

EXERCISES XV.

1. Prove that if M has property S, so also has each set M_0 satisfying the relation $M \subset M_0 \subset \mathrm{Cl}(M)$.

For the following exercises, consider the diagram:

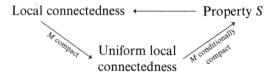

Local connectedness $\longleftarrow$ Property S

M compact   Uniform local M conditionally compact

connectedness

2. Prove the indicated implications.

3. Decide which of the implications are reversible.

Validate your conclusions with proofs or counterexamples.

48

SOLUTIONS.

1. Prove that if M has property S, so also has each set M_0 satisfying the relation $M \subset M_0 \subset \text{Cl}(M)$.

We first note that diam $M = $ diam $\text{Cl}(M)$. Since $M \subset \text{Cl}(M)$, diam $M \leq $ diam $\text{Cl}(M)$. Let $\varepsilon > 0$ be given, and let $x, y \in \text{Cl}(M)$. Then there exist points $x', y' \in M$ such that $\rho(x,x') < \varepsilon/2$ and $\rho(y,y') < \varepsilon/2$. Then

$$\rho(x,y) \leq \rho(x,x') + \rho(x',y') + \rho(y',y) \leq \varepsilon + \text{diam } M$$

Since x and y were arbitrary, as was ε, we have diam $\text{Cl}(M) \leq$ diam M.

Now suppose $\varepsilon > 0$ is given and $M = \bigcup_1^N M_i$, where each M_i is connected and has diameter less than ε. Then $M_0 = \bigcup_1^N \text{Cl}(M_i) \cap M_0$, where $\text{Cl}(M_i) \cap M_0$ is connected for each i (since $M_i \subset \text{Cl}(M_i) \cap M_0 \subset \text{Cl}(M_i)$) and diam $\text{Cl}(M_i) \cap M_0$ is less than ε. Hence M_0 has property S.

2. Prove the indicated implications.

(a) Property $S \Rightarrow$ local connectedness. Let M be a set in a metric space with property S; let $\varepsilon > 0$ be given, and choose $p \in M$ with U being the ε-sphere centered at p. Then $M = \bigcup_1^N M_i$, where each M_i is connected and has diameter less than $\varepsilon/2$. If K is the union of all M_i's satisfying $p \in \text{Cl}(M_i)$, then K is connected. Also, if $x, y \in K$, then $\rho(x,y) \leq \rho(x,p) + \rho(p,y)$, and since $x \in M_i$, $y \in M_j$, and $p \in \text{Cl}(M_i) \cap \text{Cl}(M_j)$ for some i and j, we have $\rho(x,y) < \varepsilon/2 + \varepsilon/2 = \varepsilon$. Thus the diameter of K is less than ε. Since p cannot be a limit point of K^c, there is an open set V containing p and completely contained in K. Then the open set $U \cap V$ satisfies $p \in U \cap V \subset U$ and $(U \cap V) \cap M \subset K$, where K is a connected subset of U. Hence M is locally connected.

(b) Local connectedness, compactness $\Rightarrow$ uniform local connectedness. Let M be compact and locally connected in a metric space, and let $\varepsilon > 0$ be given. We can cover M with regions in M of diameter less than $\varepsilon/2$, so that using the compactness of M, we have $M = \bigcup_1^N R_i$, where each R_i is open in M, is connected, and has diameter less than $\varepsilon/2$. For each pair R_i, R_j satisfying $\text{Cl}(R_i) \cap \text{Cl}(R_j) = \varnothing$, let $\delta_{ij} = \rho(R_i,R_j)$, so that each $\delta_{ij} > 0$. Define δ by $\delta = \min[\delta_{ij}]$ if $[\delta_{ij}] \neq 0$; otherwise, $\delta = $ diam M. Now choose $x, y \in M$ such that $\rho(x,y) < \delta$. Then for some i and j, we have $x \in \text{Cl}(R_i)$, $y \in \text{Cl}(R_j)$, where $R_i \cap R_j \neq \varnothing$. The set $\text{Cl}(R_i) \cup \text{Cl}(R_j)$ then has the properties of being connected, containing x and y, and having diameter less than ε. Hence M is uniformly locally connected.

(c) Uniform local connectedness, conditional compactness $\Rightarrow$ property S. Let M be a uniformly locally connected, conditionally compact set in a metric space, and let $\varepsilon > 0$ be given. Then there exists a $\delta > 0$ such that if $\rho(x,y) < \delta$, then x and y lie in a connected subset of diameter less than $\varepsilon/2$. Since $\text{Cl}(M)$ must be compact, we know that $\text{Cl}(M)$ is separable, and hence so also is M. Let $P = [p_i]$ be a countable dense set in M. Define, for each n, the set R_n to be the set of all those points of M which lie together with p_n in a connected set of diameter less than $\varepsilon/2$. Then R_n is connected and has diameter less than ε. We claim that $M = \bigcup_1^N R_i$ for some N. Suppose not; then there exists an infinite subsequence $[p_{n_i}]$ of $[p_i]$ such that for each i, p_{n_i} is not contained in $\bigcup_1^{n_i-1} R_n$. Since M is conditionally compact, $[p_{n_i}]$ has a limit point p. Then for some n_i and n_j $(i > j)$, $\rho(p_{n_i},p_{n_j}) < \delta$, so that $p_{n_i} \in R_{n_j} \subset \bigcup_1^{n_i-1} R_n$. But this contradicts the construction of $[p_{n_i}]$. Hence $M = \bigcup_1^N R_i$ for some N, and M has property S.

3. Decide which of the implications are reversible.

(a) Uniform local connectedness $\Rightarrow$ local connectedness. Let M be a uniformly locally connected set in a metric space. Let $p \in M$, and let U be an ε-sphere centered at p, where $\varepsilon > 0$ is arbitrary. Since M is uniformly locally connected, there exists an $\varepsilon' > 0$ such that if $\rho(x, y) < \varepsilon'$, then x and y lie together in a connected subset of M of diameter less than $\varepsilon/2$. Consider $V_\delta = V_\delta(p)$, where $\delta = \min\{\varepsilon, \varepsilon'\}$. If $x \in V \cap M$, then x and p lie together in a connected subset of M of diameter less than $\varepsilon/2$; call this subset K_x. Then $\bigcup_{x \in V \cap M} K_x$ is connected and is contained in U. Therefore, since $V \cap M \subset \bigcup K_x$, it follows that $V \cap M$ lies in a connected subset of $U \cap M$ and that $p \in V \subset U$. Hence M is locally connected.

(b) Property S, conditional compactness $\Rightarrow$ uniform local connectedness. Consider $M = [0,1) \cup (1,2]$.

(c) Local connectedness $\Rightarrow$ property S. Consider $M = \mathbb{R}$.

Functions and Mappings

Definition. A function $f: X \to Y$ is *open* (*closed*) if and only if the image of each open (closed) set in X is open (closed) in Y.

Definition. A function is *quasicompact* if and only if the image of every open inverse set is open. [A set satisfying $f^{-1}(f(A)) = A$ is called an *inverse set*.]

Remark. All open (closed) onto functions are quasicompact.

Definition. If $f: X \to Y$ is a 1-1 onto continuous function and f^{-1} is also continuous, then f is called a *homeomorphism* (or a *topological transformation*).

Definition. If X and Y are metric spaces, a function $f: X \to Y$ is *uniformly continuous* if and only if for every $\varepsilon > 0$ there exists a $\delta > 0$ such that if $x, y \in X$ and $\rho(x, y) < \delta$, then $\rho(f(x), f(y)) < \varepsilon$.

EXERCISES XVI.

1. If $f: X \to Y$, $g: Y \to Z$ are mappings, so also is $gf: X \to Z$.

2. Compactness is invariant under all mappings; i.e., if $f: X \to Y$ is a mapping and X is compact, then $f(X)$ is compact.

3. Connectedness is invariant under all mappings.

4. If X and Y are Hausdorff spaces and X is compact, then all mappings $f: X \to Y$ are closed.

5. *Example*: Local connectedness is not invariant under all mappings.

6. Local connectedness is invariant under all quasicompact mappings.

7. Any 1-1 onto quasicompact mapping is a homeomorphism.

8. If X and Y are Hausdorff spaces and X is compact, then any 1-1 onto mapping $f : X \to Y$ is a homeomorphism.

9. On a compact metric space, any continuous function to a metric space is uniformly continuous.

10. Property S is invariant under all uniformly continuous mappings.

11. *Examples*: Neither Property S nor uniform local connectedness is a topological property invariant under homeomorphisms.

SOLUTIONS.

1. If $f: X \to Y$, $g: Y \to Z$ are mappings, so also is $gf: X \to Z$.

It is sufficient to show that if U is open in Z, then $(gf)^{-1}(U)$ is open in X. But by the continuity of g and f, $g^{-1}(U)$ is open in Y and $f^{-1}(g^{-1}(U)) = (gf)^{-1}(U)$ is open in X.

2. Compactness is invariant under all mappings; i.e., if $f: X \to Y$ is a mapping and X is compact, then $f(X)$ is compact.

Let $f: X \to Y$ be a mapping, where X is compact. We wish to show that $f(X)$ is compact, so let $[U_\alpha]$ be an open cover for $f(X)$. Then by continuity, $[f^{-1}(U_\alpha)]$ is an open cover for X, so that using compactness gives $X \subset \bigcup_1^N f^{-1}(U_i)$. But then $f(X) \subset \bigcup_1^N U_i$ and $f(X)$ is compact.

3. Connectedness is invariant under all mappings.

Let $f: X \to Y$ be a mapping, and suppose that X is connected. We wish to show that $f(X)$ is connected. Suppose not; i.e., suppose $f(X) = A \cup B$ is a separation for $f(X)$. Then $X = f^{-1}(A) \cup f^{-1}(B)$ is a separation for X, but X is connected. Hence $f(X)$ must be connected.

4. If X and Y are Hausdorff spaces and X is compact, then all mappings $f: X \to Y$ are closed.

Consider a mapping $f: X \to Y$, and let A be a closed subset of X. Then A is compact, so that $f(A)$ is compact, and hence closed.

5. *Example*: Local connectedness is not invariant under all mappings.

Let C be the point set in the plane described by

$$C = \{(0, y): -1 \leq y \leq 1\} \cup \{(x, \sin 1/x): 0 < x \leq 1\}.$$

Let X be the set C with the discrete topology; i.e., each point is an open set. Let Y be the set C with the relative topology of the plane. If $f: X \to Y$ is the identity, then we have that X is locally connected but $f(X) = Y$ is not locally connected.

6. Local connectedness is invariant under all quasicompact mappings.

Consider the spaces X and Y, where X is locally connected, and suppose that $f: X \to Y$ is onto and quasicompact. Then we wish to show that Y is locally connected by showing that a component of an open set in Y is itself open in Y. Let U be an open set in Y, and K a component of U. By continuity, $f^{-1}(U)$ is open in X and $f^{-1}(K) \subset f^{-1}(U)$. For each $x \in f^{-1}(K)$, let K_x be the component in $f^{-1}(U)$ containing x. Then K_x is open, and $f^{-1}(K) \subset \bigcup_{x \in f^{-1}(K)} K_x$. Since $f(K_x)$ is connected, we have $f(K_x) \subset K$, so that $K_x \subset f^{-1}(K)$ for each x. Hence $\bigcup K_x \subset f^{-1}(K)$, which implies that $f^{-1}(K) = \bigcup K_x$ and $f^{-1}(K)$ is an open set in X. Also $f^{-1}(K)$ is an inverse set. Thus by quasicompactness, $f(f^{-1}(K)) = K$ is open in Y, and Y is locally connected.

7. Any 1-1 onto quasicompact mapping is a homeomorphism.

Let $f: X \to Y$ be a 1-1 onto quasicompact mapping. To show that f is a homeomorphism we must show that f^{-1} is continuous, or equivalently that $f(U)$ is open in Y if U is open in X. But since f is 1-1, every set is an inverse set, so by quasicompactness the result follows.

8. If X and Y are Hausdorff spaces and X is compact, then any 1-1 onto mapping $f: X \to Y$ is a homeomorphism.

It is sufficient to show that f is an open map. If U is an open set in X, then U^c is closed, and since f is a closed map, $f(U^c)$ is also closed. Since f is 1-1 $[f(U)]^c = f(U^c)$, so that $f(U)$ is open in Y. Hence f is a homeomorphism.

9. On a compact metric space, any continuous function to a metric space is uniformly continuous.

Let X and Y be metric spaces with X compact, and let $f : X \to Y$ be a mapping. Suppose that f is not uniformly continuous. Then there exists an $\varepsilon > 0$ and two sequences $\{x_n\}$ and $\{y_n\}$ of points in X such that $\rho(x_n, y_n) < 1/n$ but $\rho(f(x_n), f(y_n)) \geq \varepsilon$. We can assume that $\{x_n\}$ is an infinite set, and since X is compact, we can also assume that $x_n \to x \in X$. But then $y_n \to x$, so that by continuity, $f(x_n) \to f(x)$ and $f(y_n) \to f(x)$. Hence for large n, $\rho(f(x_n), f(y_n)) < \varepsilon$, which contradicts our choice of the two sequences. Thus f must be uniformly continuous.

10. Property S is invariant under all uniformly continuous mappings.

Consider the metric spaces X and Y, where X has the property S and $f : X \to Y$ is uniformly continuous and onto Y. Let $\varepsilon > 0$ be given. There exists a $\delta > 0$ such that if $\rho(x, y) < \delta$, then $\rho(f(x), f(y)) < \varepsilon$. Since X has property S, $X = \bigcup_1^N M_i$, where each M_i is connected and has diameter less than δ. Then $Y = \bigcup_1^N f(M_i)$, where each $f(M_i)$ is connected and has diameter less than ε. Hence Y has property S.

11. *Examples*: Neither property S nor uniform local connectedness is a topological property invariant under homeomorphisms.

Consider the following spaces and homeomorphisms:

(i) $X = (0,1]$, $Y = [1, \infty)$; $f : X \to Y$ defined by

$$f(x) = 1/x.$$

(ii) $X = [-2, -1) \cup (1, 2]$, $Y = [-1, 0) \cup (0, 1]$; $f : X \to Y$ defined by

$$f(x) = \begin{cases} x + 1 & \text{if } x \in [-2, -1), \\ x - 1 & \text{if } x \in (1, 2]. \end{cases}$$

Complete Spaces

Definition. A sequence $\{x_n\}$ in a metric space is said to be a *Cauchy sequence* if for every $\varepsilon > 0$ there exists an integer N such that if $n, m > N$, then $\rho(x_n, x_m) < \varepsilon$.

Definition. A metric space X is *complete* if and only if every Cauchy sequence of points in X converges to a point in X.

Definition. A space X is topologically complete if and only if it is homeomorphic to a complete space.

Definition. If X is any metric space, the *complete enclosure* $\tilde{X}$ of X is the space whose points are equivalence classes of Cauchy sequences $(x_1, x_2, \ldots)$ in X, where $(x_1, x_2, \ldots)$ and $(y_1, y_2, \ldots)$ are equivalent if and only if $(x_1, y_1, x_2, y_2, \ldots)$ is a Cauchy sequence, and in which the distance $\rho(x, y)$ is defined by

$$\rho(x, y) = \lim_{n \to \infty} \rho(x_n, y_n)$$

if $x = (x_1, x_2, \ldots)$ and $y = (y_1, y_2, \ldots)$. Any point $x \in X$ is identified with the point $(x, x, x, \ldots) \in \tilde{X}$, so that X is imbedded in $\tilde{X}$.

Definition. A metric space X is *totally bounded* if and only if for every $\varepsilon > 0$, $X = \bigcup_1^N X_i$, where diam $X_i < \varepsilon$.

Definition. If X and Y are metric spaces and Y is bounded, then the *function space* Y^X is the set of all mappings of X into Y. For $f, g \in Y^X$, $\rho(f, g) = \sup_{x \in X} \rho(f(x), g(x))$.

EXERCISES XVII.

1. Every compact metric space is complete. Every closed subset of a complete space is itself complete.

2. Every totally bounded complete set is compact.

3. The complete enclosure $\tilde{X}$ of any metric space X is a complete space containing X isometrically (i.e., the distance between points in X doesn't change from X to $\tilde{X}$), and X is dense in $\tilde{X}$.

4. A metric space is complete if and only if $X = \tilde{X}$.

5. Given a mapping $f: D \to Y$ where $D \subset X$, X and Y are metric spaces, Y complete, and f is uniformly continuous, then there exists a unique uniformly continuous extension of f to $Cl(D)$.

6. If X and Y are metric spaces, and Y is bounded and complete, then the mapping space Y^X is also complete. *Hint*: Prove the following lemmas.

Lemma 1. *If $\{f_n\}$ is a Cauchy sequence in Y^X and for each $x \in X$ we define $f(x) = \lim_{n \to \infty} f_n(x)$, then the sequence of functions $\{f_n\}$ converges uniformly to the function f.*

Lemma 2. *If the sequence of functions $\{f_n\}$ in Y^X converges uniformly to $f: X \to Y$, then f is continuous.*

SOLUTIONS.

1. Every compact metric space is complete. Every closed subset of a complete space is itself complete.

Let X be a compact metric space, and let $\{x_n\}$ be a Cauchy sequence in X. If some x_i is repeated an infinite number of times, then $x_n \to x_i$. Suppose on the other hand that no x_i is repeated more than a finite number of times. Then $\{x_n\}$ is an infinite set, and since X is compact, $\{x_n\}$ must have a limit point x. But then $x_n \to x$, so that in both cases the sequence converges to a point in X, and X is complete.

Now suppose that Y is a complete metric space and A is a closed subset of Y. Let $\{y_n\}$ be a Cauchy sequence in A. Then $y_n \to y \in Y$, since Y is complete; but then y is a limit point of the set $\{y_n\}$ or y is y_n for all but a finite number of n, so that $y \in A$. Hence A is complete.

2. Every totally bounded complete set is compact.

Let K be a totally bounded complete set. By the same method used to show that every compact metric space is separable, we see that every totally bounded space is separable. Hence compactness in K is equivalent to K being a B-W set, and we will show that K is a B-W set. Let A be an infinite set in K. Let $\varepsilon = 1$. Then $K = \bigcup_1^N K_{1i}$, where the diameter of K_{1i} is less than 1 for each i. Then at least one of the K_{1i}'s must contain an infinite number of points of A. We can assume that K_{11} satisfies this; i.e., $K_{11} \cap A = A_1$ is infinite. Pick $x_1 \in A_1$. Continuing, for each n, let $\varepsilon = 1/n$, and then $K = \bigcup_1^{N_n} K_{ni}$, where the diameter of K_{ni} is less than $1/n$ for each i. At least one K_{ni} contains an infinite number of points of A_{n-1}. Assuming that $K_{nn} \cap A_{n-1} = A_n$ is infinite, choose $x_n \in A_n$.

In this manner we obtain an infinite sequence $\{x_n\}$ of points of K. Let $\varepsilon > 0$ be given, and let M be an integer such that $1/M < \varepsilon$. If $n, m > M$, then x_n and x_m are contained in $A_m = K_{MM} \cap A_{M-1}$, where the diameter of K_{MM} is less than $1/M < \varepsilon$. Thus $\rho(x_m, x_n) < \varepsilon$, and $\{x_n\}$ is a Cauchy sequence. Since K is complete, we have $x_n \to x \in K$, so that K is a B-W set and hence compact.

3. The complete enclosure $\tilde{X}$ of any metric space X is a complete space containing X isometrically (i.e., the distance between points in X doesn't change from X to $\tilde{X}$), and X is dense in $\tilde{X}$.

Given that $x = (x_1, x_2, \ldots)$ and $y = (y_1, y_2, \ldots)$ are in the same equivalence class, we have that $\lim_{n \to \infty} \rho(x_n, y_n) = 0$. Hence by continuity of the metric in X, we see that the metric in $\tilde{X}$ is well defined. It follows immediately that $\tilde{X}$ is a metric space. Also, if x, $y \in X$, let $\tilde{x} = (x, x, \ldots)$ and $\tilde{y} = (y, y, \ldots) \in \tilde{X}$, and we have $\rho(\tilde{x}, \tilde{y}) = \lim_{n \to \infty} \rho(x, y) = \rho(x, y)$, so that X is imbedded isometrically in $\tilde{X}$.

To see that X is dense in $\tilde{X}$, let $\tilde{x} \in \tilde{X}$, and suppose that $\tilde{x} \notin X$. Then $\tilde{x}$ can be represented by a Cauchy sequence $(x_1, x_2, \ldots)$ where $x_i \in X$. Let $\varepsilon > 0$ be given, and let $\tilde{x}_i = (x_i, x_i, x_i, \ldots) \in \tilde{X}$ for each i. Since $(x_1, x_2, \ldots)$ is Cauchy, there exists an integer N such that $\rho(x_i, x_j) < \varepsilon$ for $i, j \geq N$. Then $\rho(\tilde{x}, \tilde{x}_N) = \lim_{n \to \infty} \rho(x_n, x_N) \leq \varepsilon$, so that $\tilde{x}$ is a limit point of X and X is dense in $\tilde{X}$.

Now let $\{\tilde{x}_n\}$ be a Cauchy sequence in $\tilde{X}$, where $\tilde{x}_n$ is represented by $(x_{n1}, x_{n2}, \ldots)$. We can assume that the diameter of the set $\{x_{n1}, x_{n2}, \ldots\}$ is less than $1/n$, since for some K, $\rho(x_{ni}, x_{nj}) < 1/n$ for $i, j \geq K$. Then $(x_{n1}, x_{n2}, \ldots)$ is equivalent to $(x_{nK}, x_{nK+1}, \ldots)$ and $\tilde{x}_n$ can be represented by the latter. We claim that $(x_{11}, x_{22}, x_{33}, \ldots)$ is a Cauchy sequence. If $\varepsilon > 0$ is given, then there exists an integer N such that $\rho(\tilde{x}_m, \tilde{x}_n) = \lim_{K \to \infty} \rho(x_{mK}, x_{nK}) < \varepsilon/3$ for $m, n \geq N$, which implies that for some fixed $K \geq N$, $\rho(x_{mK}, x_{nK}) < \varepsilon/3$ for $m, n \geq N$. Now assume that $1/N < \varepsilon/3$. Then for $m, n \geq N$

we have $\rho(x_{mm},x_{nn}) \leq \rho(x_{mm},x_{mK}) + \rho(x_{mK},x_{nK}) + \rho(x_{nK},x_{nn}) < 3 \cdot \varepsilon/3 = \varepsilon$. Hence $(x_{11}, x_{22}, \ldots)$ is a Cauchy sequence; if we define $\tilde{x}$ to be that point in X represented by $(x_{11}, x_{22}, \ldots)$, then $\tilde{x}_m \to \tilde{x}$, since $\rho(\tilde{x}_n,\tilde{x}) = \lim_{K \to \infty} \rho(x_{nK},x_{KK}) < \varepsilon/3$ for $n \geq N$. Therefore X is complete.

4. A metric space X is complete if and only if $X = \tilde{X}$.

Clearly if $X = \tilde{X}$, then X is complete. Assume then that X is complete, and let $\tilde{x} \in \tilde{X}$, where $\tilde{x}$ is represented by the Cauchy sequence $(x_1, x_2, \ldots)$. Since $x_n \to x \in X$ by the completeness of X, we have that $(x_1, x_2, \ldots)$ is equivalent to $(x, x, x, \ldots)$, so that $\tilde{x}$ can be represented by $(x, x, \ldots)$, and $\tilde{x} \in X$. Then $\tilde{X} \subset X$; hence $X = \tilde{X}$.

5. Given a mapping $f: D \to Y$ where $D \subset X$, X and Y are metric spaces, Y is complete, and f is uniformly continuous, then there exists a unique uniformly continuous extension of f to $\text{Cl}(D)$.

If $x \in \text{Cl}(D)$, let $\{x_n\}$ be a sequence in D such that $x_n \to x$. If $x \in D$, let this sequence be $(x, x, \ldots)$. Then consider the sequence $\{f(x_n)\}$ in Y. Let $\varepsilon > 0$ be given, and let $\delta > 0$ be such that if $\rho(x,y) < \delta$, then $\rho(f(x),f(y)) < \varepsilon$. If N is the integer such that for m, $n \geq N$, $\rho(x_n,x_m) < \delta$, then $\rho(f(x_n),f(x_m)) < \varepsilon$. Hence $\{f(x_n)\}$ is a Cauchy sequence in the complete space Y, so we can define $f(x) = \lim_{n \to \infty} f(x_n)$.

Suppose $\{y_n\}$ is another sequence converging to the point x. Then we have $\{f(y_n)\} \to y \in Y$. But since $\rho(x_n,y_n) \to 0$, we also have $\rho(f(x_n),f(y_n)) \to 0$, so that $y = f(x)$. Hence the definition of $f(x)$ does not depend on the choice of the sequence $\{x_n\}$.

Let $\varepsilon > 0$ be given, and let $\delta > 0$ be such that if $x, y \in D$, $\rho(x,y) < \delta$, then $\rho(f(x),f(y)) < \varepsilon$. Now suppose that $x, y \in \text{Cl}(D)$ such that $\rho(x,y) < \delta$, and $\{x_n\}$, $\{y_n\}$ are the sequences given above corresponding to these points. Then for some integer N, $\rho(x_n,y_n) < \delta$ for all $n > N$, so that $\rho(f(x_n),f(x_m)) < \varepsilon$ for $n > N$; hence $\rho(f(x),f(y)) \leq \varepsilon$. Thus f is uniformly continuous on $\text{Cl}(D)$. That f is a uniquely determined extension on $\text{Cl}(D)$ follows from the fact that D is dense in $\text{Cl}(D)$.

6. If X and Y are metric spaces, and Y is bounded and complete, then the mapping space Y^X is also complete.

Lemma 1. *If $\{f_n\}$ is a Cauchy sequence in Y^X and for each $x \in X$ we define $f(x) = \lim_{n \to \infty} f_n(x)$, then the sequence of functions converges uniformly to the function f.*

It is clear that since $\{f_n\}$ is a Cauchy sequence in Y^X, $\{f_n(x)\}$ is a Cauchy sequence in Y for each x. Hence $\lim_{n \to \infty} f_n(x)$ exists for all $x \in X$, and f is defined on X.

Now let $\varepsilon > 0$ be given. Then there exists an integer N such that for $m, n \geq N$, $\rho(f_n,f_m) < \varepsilon$. Letting $m \to \infty$, we have $\rho(f_n,f) \leq \varepsilon$, and the sequence of functions $\{f_n\}$ converges to f uniformly.

Lemma 2. *If the sequence of functions $\{f_n\}$ in Y^X converges uniformly to $f: X \to Y$, then f is continuous.*

Let $x_0 \in X$, and let $\varepsilon > 0$ be given. Then there exists an integer N such that $\rho(f_n(x),f(x)) < \varepsilon/3$ for $n \geq N$ and all $x \in X$. Let U be an open set containing x_0 such that for all $x \in U$, $\rho(f_n(x),f_n(x_0)) < \varepsilon/3$, by the continuity of f_n. Then we have

$$\rho(f(x),f(x_0)) \leq \rho(f(x),f_n(x)) + \rho(f_n(x),f_n(x_0)) + \rho(f_n(x_0),f(x_0))$$
$$< \varepsilon/3 + \varepsilon/3 + \varepsilon/3 = \varepsilon$$

for all $x \in U$. Hence f is continuous, since x_0 was arbitrary.

First Semester Examination

1. Assuming that the real interval $I = [0,1]$ is compact, why is it connected?

2. Verify that in Q_ω or Q'_ω a sequence of points $\{x_n\}$ converges to a point $p = (p_1, p_2, \ldots)$ if and only if for each m, the mth coordinates $x_{1m}, x_{2m}, \ldots$ of $x_1, x_2, \ldots$ respectively converge to p_m.

   Verify that a sequence of points x_n in H converges to a point $p = (p_1, p_2, \ldots)$ in H if and only if $\|x_n\| \to \|p\|$ and $x_{nk} \to p_k$ for each k.

Table 1

Space	E^n	H	Q_ω	Q'_ω	H_1: $x_1 = 1$	S: $\|x\| = 1$	I^n
Explanation	Euclidean n-space	Hilbert space			"Plane" $x_1 = 1$ in H	Unit sphere in H	$I \times I \times \cdots \times I$
Compact							
Locally compact							
Convex							
Connected							
Locally connected							
Unif. loc. connected							
Complete							

3. Fill in Table 1, entering $+$ if true, $-$ if false.

4. Prove that H and H_1 are homeomorphic.

5. Prove that Q_ω and Q'_ω are homeomorphic.

6. Prove that H_1 is topologically embeddable in S (i.e., homeomorphic to a subset of S).

Solutions:

1. Assuming that the real interval $I = [0,1]$ is compact, why is it connected?

Suppose that $x, y \in I$, and assume that $x < y$. Let $\varepsilon > 0$ be given. If $y - x = \delta$, then let N be an integer satisfying $\delta/N < \varepsilon$. Then we have that $x, x + \delta/N, x + 2\delta/N, \ldots, x + (N - 1)\delta/N, y$ is an ε-chain joining x and y. Hence I is a compact, well-chained set, so that I is connected.

2. Verify that in Q_ω and Q'_ω a sequence of points $\{x_n\}$ converges to a point $p = (p_1, p_2, \ldots)$ if and only if for each m, the mth coordinates $x_{1m}, x_{2m}, \ldots$ of $x_1, x_2, \ldots$ respectively converge to p_m.

(a) Q_ω: suppose $\{x_n\}$ converges to p, and let $\varepsilon > 0$ be given. Then there exists an integer N such that $\rho(x_n, p) < \varepsilon$ for $n \geq N$; i.e., $\sqrt{\sum_1^\infty (x_{nm} - p_m)^2} < \varepsilon$. But then for each m

$$|x_{nm} - p_m| = \sqrt{(x_{nm} - p_m)^2} \leq \sqrt{\sum_1^\infty (x_{nm} - p_m)^2}$$

for all $n \geq N$, so that $x_{nm} \to p_m$ for each m.

Now suppose that for each m, the mth coordinates converge to p_m. Let $\varepsilon > 0$ be given, and let N be an integer such that $\sqrt{\sum_N^\infty 1/m^2} < \varepsilon/2$. Then by the restriction on each coordinate we have $\sqrt{\sum_{m=N}^\infty (x_{nm} - p_m)^2} < \varepsilon/2$ for all n. Now using the fact that each coordinate converges, let N_m be an integer such that $|x_{nm} - p_m| < \varepsilon/2N$ for all $n \geq N_m$, $m = 1, 2, \ldots, N - 1$. If we let $N^* = \max(N_1, N_2, N_3, \ldots, N_{N-1})$, then we have for $n \geq N^*$

$$\sqrt{\sum_1^\infty (x_{nm} - p_m)^2} \leq \sqrt{\sum_1^{N-1} (x_{nm} - p_m)^2} + \sqrt{\sum_N^\infty (x_{nm} - p_m)^2}$$

$$< \sum_1^{N-1} |x_{nm} - p_m| + \frac{\varepsilon}{2}$$

$$< \sum_1^{N-1} \frac{\varepsilon}{2N} + \frac{\varepsilon}{2} < \frac{\varepsilon}{2} + \frac{\varepsilon}{2} = \varepsilon.$$

Hence $x_n \to p$.

(b) Q'_ω: Suppose that $x_n \to p$; let $\varepsilon > 0$ be given, and consider the mth coordinate. There exists an integer N such that

$$\sum_1^\infty \frac{|x_{km} - p_m|}{2^m} < \frac{\varepsilon}{2^m} \quad \text{for } k \geq N.$$

Then

$$|x_{km} - p_m| \, 2^{-m} \leq \sum_1^\infty |x_{km} - p| \, 2^{-m} < \frac{\varepsilon}{2^m}$$

and

$$|x_{km} - p_m| < \varepsilon$$

for $k \geq N$. Hence $x_{km} \to p_m$.

Now suppose that for each m, $x_{km} \to p_m$. Let N be an integer such that $\sum_N^\infty 2^{-m} < \varepsilon/2$, where $\varepsilon > 0$ is given. Then $\sum_{m=N}^\infty |x_{km} - p_m| \, 2^{-m} < \varepsilon/2$. By convergence of the coordinates, we have integers K_m such that for $k \geq K_m$, $|x_{km} - p_m| < \varepsilon/2N$ $(m = 1, 2, \ldots, N - 1)$.

Then we have

$$\sum_{1}^{\infty} \frac{|x_{km} - p_m|}{2^m} = \sum_{m=1}^{N-1} \frac{|x_{km} - p_m|}{2^m} + \sum_{m=N}^{\infty} \frac{|x_{km} - p_m|}{2^m}$$

$$\leq \sum_{1}^{N-1} \frac{\varepsilon}{2N} + \frac{\varepsilon}{2} < \varepsilon$$

for $k \geq \max(K_1, K_2, \ldots, K_{N-1})$ and $x_k \to p$.

Verify that a sequence of points $\{x_n\}$ in H converges to a point $p = (p_1, p_2, \ldots)$ in H if and only if $\|x_n\| \to \|p\|$ and $x_{nk} \to p_k$ for each k.

Suppose that $x_n \to p$. Applying the same method used in (a) and (b), we see that $x_{nk} \to p_k$ for each k. If we let $0 = (0, 0, \ldots)$, then using the inequalities $\rho(x_n,0) \leq \rho(x_n,p) + \rho(p,0)$ and $\rho(p,0) \leq \rho(p,x_n) + \rho(x_n,0)$, we have that $|\rho(p,0) - \rho(x_n,0)| \leq \rho(x_n,p)$, or equivalently, $|\,\|x_n\| - \|p\|\,| \leq \rho(x_n,p)$. Since $x_n \to p$, $\|x_n\| \to \|p\|$.

Now suppose that $\|x_n\| \to \|p\|$ and $x_{nk} \to p_k$ for each k. Define $q(n)$ for $n = 1, 2, \ldots$ by

$$q(n) = \|x_n\|^2 - \|p\|^2.$$

If $\varepsilon > 0$ is given, choose an integer $M > 0$ such that

$$\sum_{M+1}^{\infty} p_k^2 < \frac{\varepsilon^2}{12}.$$

Now define

$$r(n) = \sum_{k=1}^{M} (p_k^2 - x_{nk}^2),$$

$$s(n) = \sum_{k=1}^{M} p_k(p_k - x_{nk})$$

for $n = 1, 2, \ldots$. Then $\lim_{n \to \infty} r(n) = \lim_{n \to \infty} s(n) = \lim_{n \to \infty} q(n) = 0$. Hence there must exist an integer N such that for $n \geq N$

$$|r(n)| + |s(n)| + |q(n)| < \varepsilon^2/12.$$

We note here that

$$\sum_{k=M+1}^{\infty} x_{nk}^2 = \|x_n\|^2 - \sum_{k=1}^{M} x_{nk}^2$$

$$= \|x_n\|^2 - \|p\|^2 + \sum_{k=1}^{M} (p_k^2 - x_{nk}^2) + \sum_{M+1}^{\infty} p_k^2$$

$$= q(n) + r(n) + \sum_{M+1}^{\infty} p_k^2.$$

Then we have for $n \geq N$

$$\sum_{k=M+1}^{\infty} x_{nk}^2 \leq \frac{\varepsilon^2}{12} + \frac{\varepsilon^2}{12} = \frac{\varepsilon^2}{6}.$$

Now since

$$\sum_{k=1}^{\infty} (x_{nk} - p_k)^2 = \|x_n\|^2 + \|p\|^2 - 2\sum_1^{\infty} p_k x_{nk}$$

$$= \|x_n\|^2 - \|p\|^2 + 2\sum_1^{\infty} p_k(p_k - x_{nk})$$

$$= q(n) + 2s(n) + 2\sum_{M+1}^{\infty} p_k^2 - 2\sum_{M+1}^{\infty} p_k x_{nk}$$

if $n \geq N$, we have

$$\sum_1^{\infty} (x_{nk} - p_k)^2 < \frac{\varepsilon^2}{12} + \frac{\varepsilon^2}{6} + \frac{\varepsilon^2}{6} + 2\sum_{M+1}^{\infty} |p_k x_{nk}|$$

$$\leq \frac{5\varepsilon^2}{12} + 2\sqrt{\sum_{M+1}^{\infty} p_k^2} \sqrt{\sum_{M+1}^{\infty} x_{nk}^2}$$

$$\leq \frac{5\varepsilon^2}{12} + 2\sqrt{\frac{\varepsilon^2}{12}}\sqrt{\frac{\varepsilon^2}{6}} < \frac{3\varepsilon^2}{4}.$$

In other words, for $n \geq N$, $\rho(x_n, p) < \sqrt{3\varepsilon^2/4} < \varepsilon$ and $x_n \to p$.

3. Fill in Table 1, entering $+$ if true, $-$ if false.

Filled-in Table 1 is given in Table 2.

Table 2

Space	E^n	H	Q_ω	Q'_ω	H_1: $x_1 = 1$	S: $\|x\| = 1$	I^n
Explanation	Euclidean n-space	Hilbert space			"Plane" $x_1 = 1$ in H	Unit sphere in H	$I \times I \times \cdots \times I$
Compact	$-$	$-$	$+$	$+$	$-$	$-$	$+$
Locally compact	$+$	$-$	$+$	$+$	$-$	$-$	$+$
Convex	$+$	$+$	$+$	$+$	$+$	$-$	$+$
Connected	$+$	$+$	$+$	$+$	$+$	$+$	$+$
Locally connected	$+$	$+$	$+$	$+$	$+$	$+$	$+$
Unif. loc. connected	$+$	$+$	$+$	$+$	$+$	$+$	$+$
Complete	$+$	$+$	$+$	$+$	$+$	$+$	$+$

Proposition. *H is complete.*

To prove this we make use of the following lemma:

Lemma. *H is linear; i.e., if x, y ∈ H, then ax + by ∈ H for any real numbers a and b.*

PROOF. It is sufficient to show that $x + y \in H$, since clearly if $x \in H$, then $ax \in H$. If $x = (x_1, x_2, \ldots)$ and $y = (y_1, y_2, \ldots)$, we have

$$\left| \sum_1^M (x_n + y_n)^2 \right| = \left| \sum_1^M x_n^2 + 2 \sum_1^M x_n y_n + \sum_1^M y_n^2 \right|$$

$$\leq \sum_1^M x_n^2 + \sum_1^M y_n^2 + 2 \sum_1^M |x_n y_n|$$

$$\leq \sum_1^M x_n^2 + \sum_1^M y_n^2 + 2 \sqrt{\sum_1^M y_n^2} \sqrt{\sum_1^M y_n^2}$$

$$\leq \left[\sqrt{\sum_1^M x_n^2} + \sqrt{\sum_1^M y_n^2} \right]^2$$

$$\leq [\|x\| + \|y\|]^2.$$

Letting $M \to \infty$, we have that $\|x + y\| \leq \|x\| + \|y\| < \infty$ and $x + y \in H$. □

PROOF OF COMPLETENESS OF H. Let $\{x_n\}$ be a Cauchy sequence in H where $x_n = (x_{ni}, x_{n2}, \ldots)$. Let $p_k = \lim_{n \to \infty} x_{nk}$ (which always exists, since E is complete), and let $p = (p_1, p_2, \ldots)$. We claim $p \in H$ and $x_n \to p$. Let $\varepsilon > 0$ be given, and determine an integer N such that $\rho(x_n, x_m) < \sqrt{\varepsilon}$ for $n, m \geq N$ or $\sum_{k=1}^\infty (x_{nk} - x_{mk})^2 < \varepsilon$ for $n, m \geq N$. Therefore for each M, $\sum_1^M (x_{nk} - x_{mk})^2 < \varepsilon$ if $n, m \geq N$. Fixing $m \geq N$ and letting $n \to \infty$ gives

$$\sum_{k=1}^M (p_k - x_{mk})^2 \leq \varepsilon$$

for each M. Hence

$$\sum_{k=1}^\infty (p_k - x_{mk})^2 = \rho(p, x_m) \leq \varepsilon \quad \text{for } m \geq N,$$

and $x_m \to p$. Also, since $\sum_1^\infty (p_k - x_{mk})^2 \leq \varepsilon < \infty$, we have $p - x_m \in H$. But $x_m \in H$, so that $p - x_m + x_m = p \in H$ and H is complete. □

4. Prove that H and H_1 are homeomorphic.

Define $f : H \to H_1$ by $f(x_1, x_2, \ldots) = (1, x_1, x_2, \ldots)$. It is clear that f is 1-1 and onto H_1. Also, since f preserves distances, it follows immediately that both f and f^{-1} are continuous. Hence f is a homeomorphism.

5. Prove that Q_ω and Q'_ω are homeomorphic.

Define $f : Q_\omega \to Q'_\omega$ by $f(x_1, x_2, \ldots) = (x_1, 2x_2, \ldots, nx_n, \ldots)$. If $(x_1, x_2, \ldots) \in Q_\omega$, then $0 \leq x_n \leq 1/n$, so that $0 \leq nx_n \leq 1$ for each n and $f(Q_\omega) \subset Q'_\omega$. If $y' = (y_1, y_2, \ldots) \in Q'_\omega$, then $y = (y_1, y_2/2, \ldots, y_n/n, \ldots) \in Q_\omega$ and $f(y) = y'$, so that f is onto Q'_ω; f is clearly 1-1.

To see that f is continuous, let $\{x_n\}$ be a sequence in Q_ω where $x_n = (x_{n1}, x_{n2}, \ldots)$, and assume that $x_n \to p$, $p = (p_1, p_2, \ldots)$. For each k, $x_{nk} \to p_k$ (using Problem 2), and hence $kx_{nk} \to kp_k$ as $n \to \infty$. It then follows that if $x'_n = (x_{ni}, 2x_{n2}, \ldots, kx_{nk}, \ldots)$ and $p' = (p_1, 2p_2, \ldots, kp_k, \ldots)$, then $x'_n \to p'$ (using Problem 2). But $x'_n = f(x_n)$ and $p' = f(p)$. Hence $x_n \to p$ implies that $f(x_n) \to f(p)$ and f is continuous.

Then since f is a 1-1 mapping of a compact Hausdorff space onto a Hausdorff space, f is a homeomorphism.

6. Prove that H_1 is topologically embeddable in S (i.e., homeomorphic to a subset of S).

If $x = (1, x_2, \ldots) \in H_1$, define f on H_1 by

$$f(x) = (1/\|x\|, x_2/\|x\|, \ldots).$$

Since $\|x\| \geq 1$, division by $\|x\|$ is permissible and $\|f(x)\| = \sqrt{\sum x_i^2}/\|x\| = \|x\|/\|x\| = 1$, so that $f(x) \in S$ and $f: H_1 \to S$. It is clear that f is 1-1. To see that f is continuous, we appeal to two theorems on continuous functions from advanced calculus. We know that $\|x\|$ is a continuous function of H_1, and hence $1/\|x\|$ is also continuous, since $\|x\| \neq 0$ for any $x \in H_1$. It then follows that $f(x) = x/\|x\|$ is the product of two continuous functions on H_1 and so is itself continuous.

Now for all $y = (y_1, y_2, \ldots) \in S$ where $y_1 > 0$, defined $g(y) = (1, y_2/y_1, y_3/y_1, \ldots)$. Then if $x = (1, x_2, x_3, \ldots)$, $gf(x) = g(1/\|x\|, x_2/\|x\|, \ldots) = (1, x_2, \ldots) = x$, so that $g = f^{-1}$. We also note that $\|g(y)\| = 1/y_1 \sqrt{\sum_1^\infty y_i^2} = 1/y_1$. To prove the continuity of g, let $\{y_n\}$ be a sequence in $f(H_1)$ such that $y_n \to p \in f(H_1)$, and let $y_n = (y_{n1}, y_{n2}, \ldots)$, $p = (p_1, p_2, \ldots)$, $g(y_n) = x_n = (x_{n1}, x_{n2}, \ldots)$, $g(p) = q = (q_1, q_2, \ldots)$. We know that $y_{nk} \to p_k$ as $n \to \infty$ for each k. Then

$$\|g(y_n)\| = \frac{1}{y_{ni}} \to \frac{1}{p_1} = \|g(p)\| \quad \text{and} \quad x_{nk} = \frac{y_{nk}}{y_{n1}} \to \frac{p_k}{p_1} = q_k$$

as $n \to \infty$. Hence by Problem 2 we have $g(y_n) \to g(y)$, and $g = f^{-1}$ is continuous.

SECTION XVIII
Mapping Theorems

EXERCISES XVIII.

1. In a metric space, any nondegenerate connected set can be mapped onto the unit interval.

2. The Cantor discontinuum can be mapped onto any compact metric space.

3. A metric space M is the continuous image of the unit interval if and only if M is a locally connected continuum.

4. If A and B are locally connected, nondegenerate continua in a metric space, then there exists a mapping of A onto B.

SOLUTIONS.

1. In a metric space, any nondegenerate connected set can be mapped onto the unit interval.

Consider a set M in a metric space, and let a and b be fixed distinct points in M. We define $f: M \to I$ by

$$f(x) = \frac{\rho(x,a)}{\rho(x,a) + \rho(x,b)}.$$

Recalling that the metric ρ is continuous and noting that $\rho(x,a) + \rho(x,b) \neq 0$ for any x, we have that f is a quotient of continuous functions and hence is itself continuous. Since $f(a) = 0$, $f(b) = 1$, and $f(M)$ is connected, f must be onto I.

2. The Cantor discontinuum can be mapped onto any compact metric space.

At this point we shall review the construction of the Cantor discontinuum. Consider the unit interval $[0,1] = C_0$. We will first construct a sequence of sets C_n. Remove the segment $(\frac{1}{3},\frac{2}{3})$ from C_0 to obtain the set

$$C_1 = [0,\tfrac{1}{3}] \cup [\tfrac{2}{3},1].$$

Remove the middle third of each of these intervals to obtain the set

$$C_2 = [0,\tfrac{1}{9}] \cup [\tfrac{2}{9},\tfrac{1}{3}] \cup [\tfrac{2}{3},\tfrac{7}{9}] \cup [\tfrac{8}{9},1].$$

In general, for each n, C_n is obtained from C_{n-1} by removing the open middle third of each interval contained in C_{n-1}. We note that each C_n is the union of 2^n closed intervals, each interval having length 3^{-n}, and $C_0 \supset C_1 \supset C_2 \supset \cdots$. We now define the Cantor discontinuum C by $C = \bigcap C_n$. It follows at once that C is compact and that C contains no intervals (of positive length); hence C is totally disconnected.

Now let S be a compact metric space. Since S is totally bounded, we have $S = \bigcup_1^{N_0} S_i$, where S_i is closed and diam $S_i < 1$. We can assume that $N_0 \geq 2$. Now let $C = \bigcup_1^{N_0} C_i$, where the C_i's are closed and disjoint. To see that this can be done, let $I_1, I_2, \ldots$ be the distinct components of $I \sim C$, where $n > m$ implies diam $I_n \leq$ diam I_m. Then $I \sim [\bigcup_1^{N_0-1} I_n]$ is a union of N_0 closed disjoint intervals, and C is contained in this union. Hence if we intersect C with these intervals, we obtain the C_i's given above. We now associate S_i with C_i.

Similarly each $S_i = \bigcup_1^{N_i} S_{ij}$, where each S_{ij} is closed and diam $S_{ij} < \frac{1}{2}$. We can assume $N_i \geq 2$ for each i. Then $C_i = \bigcup_1^{N_i} C_{ij}$, where the C_{ij}'s are closed and disjoint and we associate S_{ij} and C_{ij}.

Continuing this process, we are partitioning C into closed disjoint sets whose diameters tend to zero as the number of partitions increases. This follows because we required that each set at one stage in the decomposition of S be decomposed into at least two sets at the next stage.

If $x \in C$, then x is the intersection of a unique sequence of sets in the decomposition of C; say $x = \bigcap C_{xi}$. (By a unique sequence, we mean that one set in the sequence is chosen at each stage in the decomposition.) This follows from the fact that the sets in this decomposition are disjoint and their diameters tend to zero.

Corresponding to the sequence $\{C_{xi}\}$, we have the sequence $\{S_{xi}\}$ and $\bigcap S_{xi} = x'$, since $\{S_{xi}\}$ is a monotone decreasing sequence of compact sets whose diameters approach zero. We then define $f(x) = x'$.

Since the sequence $\{C_{xi}\}$ is unique, f is single-valued, and it is clear that f is onto S. To see that f is continuous, let $x \in C$ and $\varepsilon > 0$ be given. If $\{C_{xi}\}$ is the unique sequence whose intersection is x, and N is an integer such that $1/N < \varepsilon$, let $\delta > 0$ be a real number such that if $|x - y| < \delta$, $y \in C$, then $y \in C_{xN}$. This implies that $f(y) \in S_{xN}$ and $\rho(f(x), f(y)) < 1/N < \varepsilon$. Hence f is continuous.

3. A metric space M is the continuous image of the unit interval if and only if M is a connected continuum.

Suppose first that M is the continuous image of the unit interval I; $f: I \to M$. It follows immediately that M is compact and connected. Since I is compact, f is a closed map and hence quasicompact, and local connectedness is invariant under quasi-compactness. Thus M is a locally connected continuum.

Now suppose that M is a locally connected continuum. By Appendix I at the end of this Part, we have that every point in M is interior to a locally connected continuum of diameter < 1. Covering M with interiors of such continua and using compactness, we have that M can be written as a finite union G of locally connected continua of diameter < 1. Choose distinct points a and b in M. We now claim that the elements of G can be arranged into a chain $[E_1^{(1)}, E_2^{(1)}, \ldots, E_N^{(1)}]$ from a to b; i.e.,

$$a \in E_1^{(1)}, \quad b \in E_N^{(1)}, \quad \text{and} \quad E_i^{(1)} \cap E_{i+1}^{(1)} \neq \varnothing.$$

Since the set of all points chainable to a by elements of G is open and closed in M, this set must equal M. Hence there exists a chain $C = [C_1, \ldots, C_K]$ of elements of G joining a and b. Suppose there are N elements of G not contained in C. There exists at least one of these, say H, which intersects some C_i, or else M is not connected. If $H \cap C_j \neq \varnothing$, we construct a new chain $C' = [C_1, \ldots, C_j, H, C_j, \ldots, C_k]$ joining a and b, and there are at most $M - 1$ elements of G not contained in C'. By repeating this process, we eventually arrange the elements of G into a chain from a to b. Since each link may be repeated any number of times, we may assume that this chain has 2^{V_1} links.

We next select a sequence of points

$$a = x_0^{(1)}, x_1^{(1)}, x_2^{(1)}, \ldots, x_{2^{v_2}}^{(1)} = b,$$

where $x_i^{(1)} \in E_i^{(1)} \cap E_{i+1}^{(1)}$ for $i = 1, \ldots, 2^{v_1} - 1$.

Now express $E_i^{(1)}$ as a union of a finite number of locally connected continua, each of diameter $< \frac{1}{2}$. We can assume that this finite number is 2^{v_2} for all i, and that the continua are arranged in chains from $x_i^{(1)}$ to $x_{i+1}^{(1)}$. Our chain is now in the form

$$E_1^{(2)}, E_2^{(2)}, \ldots, E_{(i-1) \cdot 2^{v(2)}}^{(2)}, E_{(i-1) \cdot 2^{v(2)}+1}^{(2)}, \ldots, E_{(i+1) \cdot 2^{v(2)}}^{(2)}, \ldots, E_{2^{v(1)+v(2)}}^{(2)},$$

where for typographical reasons we have temporarily written $v(j)$ instead of v_j. Again choose a sequence of points

$$a = x_0^{(2)}, x_1^{(2)}, \ldots, x_{2^{v(2)}}^{(2)} = x_1^{(2)}, \ldots, x_{i \cdot 2^{v(2)}}^{(2)} = x_i^{(1)}, \ldots, x_{2^{v(1)+v(2)}}^{(2)} = b,$$

where $x_i^{(2)} \in E_i^{(2)} \cap E_{i+1}^{(2)}$.

At the third stage, express the sets $E_i^{(2)}$ as a chain from $x_{i-1}^{(2)}$ to $x_i^{(2)}$ of 2^{v_3} locally connected continua, each of diameter $< 1/2^2$, and choose points $x_j^{(3)}$ as before.

Continuing this procedure we obtain for each k a sequence of points

$$a = x_0^{(k)}, x_1^{(k)}, \ldots, x_{2^{v(1)+v(2)+\cdots+v(k)}}^{(k)} = b.$$

Let t be a dyadic rational written in the form $t = i/2^{v_1 + \cdots + v_k}$ for some k. Then $f(t) = x_i^{(k)}$, $f(t)$ is single-valued by the following: If $t = i/2^{v_1 + \cdots + v_k} = j/2^{v_1 + \cdots + v_n}$, where

$k < n$, then $f(t) = x_i^{(k)}$, $f(t) = x_j^{(n)}$. But since

$$t = \frac{i \cdot 2^{v_{k+1} + \cdots + v_n}}{2^{v_1 + \cdots + v_n}},$$

we have $x_j^{(n)} = x_{i \cdot 2^{v(k+1)} + \cdots + v(n)}^{(n)} = x_{i \cdot 2^{v(k+1)} + \cdots + v(n-1)}^{(n-1)} = \cdots = x_i^{(k)}$, and f is single-valued.

Also f is uniformly continuous on the dyadic rationals D. Let $\varepsilon > 0$ be given, let k be an integer such that $1/2^k < \varepsilon/2$, and let $\delta = 1/2^{v_1 + \cdots + v_k} = 1/2^v$. If $t_1, t_2 \in D$ and $|t_1 - t_2| < \delta$, then t_1 and t_2 lie between values $(j-1)/2^v$, $j/2^v$, $(j+1)/2^v$ for some j. Suppose that t_1 lies between the first two of these values. Then $x_{j-1}^{(k)}, x_j^{(k)} \in E_j^{(k)}$, so that $f(t_1) \in E_j^{(k)}$ by the above construction. By this same procedure, regardless of which values t_2 lies between, $f(t_2) \in E_j^{(k)} \cup E_{j+1}^{(k)}$ and diam $E_j^{(k)} \cup E_{j+1}^{(k)} < \varepsilon$. Hence $\rho(f(t_1), f(t_2)) < \varepsilon$, and f is uniformly continuous.

To complete the proof, we now extend f to I. Since $f(I)$ is compact and contains $f(D)$, which is dense in M, $f(I)$ must be all of M.

4. If A and B are locally connected, nondegenerate continua in a metric space, then there exists a mapping of A onto B.

This follows immediately from Exercises 1 and 3.

SECTION XIX

Simple Arcs and
Simple Closed Curves

Definition. A set homeomorphic with a closed interval is a *simple arc*. A set homeomorphic with a circle of positive radius is a *simple closed curve*.

Definition. A point p of a connected set M is a *cut point* of M provided that $M \sim p$ is not connected.

Definition. A point p of a connected set M *separates* two points a and b in M provided that there is a separation $M \sim p = M_a \cup M_b$ with $a \in M_a$, $b \in M_b$.

Definition. The *end points* of a simple arc are the points corresponding to the ends of the interval.

Note. End points are well defined by this definition.

EXERCISES XIX.

1. A nondegenerate metric continuum T is a simple arc "from a to b" (with end points a and b) if and only if each point of T other than a and b separates a and b in T.

2. Every nondegenerate metric continuum has at least two noncut points.

3. A nondegenerate metric continuum T is a simple arc from a to b if and only if every point of $T \sim a \sim b$ is a cut point of T.

4. A nondegenerate metric continuum T is a simple closed curve if and only if it is disconnected by the removal of any two distinct points.

SOLUTIONS.

1. A nondegenerate metric continuum T is a simple arc "from a to b" (with end points a and b) if and only if each point of T other than a and b separates a and b in T.

The necessity follows immediately from the properties of the unit interval.

Now suppose that T is a continuum in a metric space such that every point other than a or b separates a and b in T. Since T is a compact metric space, T is separable, so let $P = \bigcup_1^\infty p_i$ be a countable dense set in T. We can assume that neither a nor b is contained in P and that the p_i's are distinct.

We define our homeomorphism as follows: $h(a) = 0$, $h(b) = 1$. By our assumptions we have that $T \sim p_1 = T_a(p_1) \cup T_b(p_1)$ is a separation such that $a \in T_a(p_1)$, $b \in T_b(p_1)$. Define $h(p_1) = \frac{1}{2}$.

Now consider the two sets $T_a(p_1) \cup \{p_1\} = T_{1/4}$ and $T_b(p_1) \cup \{p_1\} = T_{3/4}$. Recalling Exercise 4 in §X both of these sets are connected. Suppose q is a limit point of $T_{1/4}$. Then $q \in T_{1/4}$, since otherwise $T_b(p_1)$ is an open set in T which contains q but no point of $T_{1/4}$. Thus $T_{1/4}$ and similarly $T_{3/4}$ are closed subsets of a compact set, and hence are themselves compact. Now let x be a point in $T_{1/4}$ other than a and p_1. We know that $T \sim x = T_a(x) \cup T_b(x)$ is a separation, where $a \in T_a(x)$, $b \in T_b(x)$. Consider the representation $T_{1/4} \sim x = (T_{1/4} \cap T_a(x)) \cup (T_{1/4} \cap T_b(x))$. This will prove that x separates a and p_1 in $T_{1/4}$ if we can show $p_1 \in T_b(x)$. Suppose not; suppose that $p_1 \in T_a(x)$. Then since $p_1 \in T_{3/4}$ and $T_{3/4}$ is connected, $T_{3/4} \subset T_a(x)$. But $b \in T_{3/4}$ and $b \notin T_a(x)$, and we have a contradiction. Hence $p \in T_b(x)$, and x separates a and p_1 in $T_{1/4}$. Hence we have shown that $T_{1/4}$ and, by symmetry, $T_{3/4}$ have the same properties as T.

Let $p_{k(1/4)}$ and $p_{k(3/4)}$ be those elements of P with the least subscript which lie in $T_{1/4}$ and $T_{3/4}$ respectively. Note that one of these points must be p_2, but the other need not be p_3. We define $h(p_{k(1/4)}) = \frac{1}{4}$, $h(p_{k(3/4)}) = \frac{3}{4}$.

By the procedure above, we obtain four continua $T_{1/8}$, $T_{3/8}$, $T_{5/8}$, $T_{7/8}$ with the assumed property. Choosing points $p_{k(1/8)}$, $p_{k(3/8)}$, $p_{k(5/8)}$, $p_{k(7/8)}$ to be the point in P with the least subscript which belongs to $T_{1/8}$, $T_{3/8}$, $T_{5/8}$, $T_{7/8}$ respectively, we define $h(p_{k(i/8)}) = i/8$ for $i = 1, 3, 5, 7$. [Note that $k(i/8) > 3$ for all i.]

In general for each n we decompose T into 2^n continua $T_{1/2^n}, T_{3/2^n}, \ldots, T_{(2^n-1)/2^n}$, where only the intersection of "adjacent" ones is nonempty. By continuing this process and choosing the points $p_{k(i/2^n)}$, we construct a map h from P onto the dyadic rationals D. That h is defined on all of P follows from our requirement that $p_{k(i/2^n)}$ is the element of P with the *least* subscript which is contained in $T_{i/2^n}$.

We now claim that h is uniformly continuous. If $\varepsilon > 0$ is given, let N be an integer such that $1/2^{N-1} < \varepsilon/2$. Consider the representation $T = \bigcup T_{i/2^N}$, $i = 1, 3, \ldots, 2^N - 1$. If $\delta_{ij} = \rho(T_{i/2^N}, T_{j/2^N})$ where $T_{i/2^N} \cap T_{j/2^N} = \varnothing$, define $\delta = \min[\delta_{ij}]$. Thus if $p_1, p_2 \in P$, $\rho(p_1, p_2) < \delta$, then p_1 and p_2 either are contained in the same continuum $T_{i/2^N}$, or are contained in adjacent continua $T_{i/2^N}$, $T_{(i+2)/2^N}$. Notice that diam $h(P \cap T_{i/2^N}) = 1/2^{N-1}$ and that $h(P \cap T_{i/2^N}) \cap h(P \cap T_{(i+2)/2^N}) \neq \varnothing$, so that diam $h[P \cap (T_{i/2^N} \cup T_{(i+2)/2^N})] < \varepsilon$ and $|h(p_1) - h(p_2)| < \varepsilon$. Hence h is uniformly continuous.

If we extend h to all of T, then $h(T)$ is compact and contains the dense set $h(P)$, so $h(T) = I$. It only remains now to show that the extended map h is 1-1, since any 1-1 continuous map from a compact Hausdorff space onto a Hausdorff space is a homeomorphism. Let $x, y \in T$, $x \neq y$. We may assume that $y \in T_b(x)$, where $T \sim x = T_a(x) \cup T_b(x)$ is a separation. Then $x \in T_a(y)$. Consider the two disjoint continua $T_a(x) \cup x$ and $T_b(y) \cup y$. Since the set $T_a(y) \cap T_b(x)$ is nonempty and open in T, we can choose $z_1, z_2 \in T_a(y) \cap T_b(x)$, where $z_1, z_2 \in P$, $z_1 \neq z_2$. Then

$$T_a(x) \cup x \subset T_a(z_i) \cup z_i = S_i$$

and

$$T_b(y) \cup y \subset T_b(z_j) \cup z_j = S_j,$$

where $S_i \cap S_j = \emptyset$, for one choice of $j \neq i$, and we may assume $i = 1, j = 2$. Then

$$h(P \cap S_1) \subset [0, h(z_1)], \qquad h(P \cap S_2) \subset [h(z_2), 1],$$

where $h(z_1) < h(z_2)$, since $z_1, z_2 \in P$. Then

$$h(T_a(x) \cup x) \subset [0, h(z_1)]$$

and

$$h(T_a(y) \cup y) \subset [h(z_2), 1];$$

in particular, $h(x) \leq h(z_1) < h(z_2) \leq h(y)$, so that $h(x) \neq h(y)$.

2. Every nondegenerate metric continuum has at least two noncut points.

We prove this by contradiction. Suppose that M is a nondegenerate metric continuum having no more than one noncut point. If x is a cut point of M and $M \sim x = A \cup B$ is a separation, then at least one of these sets, say A, contains no noncut point of M. Since M is compact and metric, M has a countable dense subset $P = \bigcup p_i$. Let n_1 be the first integer such that $p_{n_1} \in A$. Then p_{n_1} is a cut point of M, and there exists a separation $M \sim p_{n_1} = E_1 \cup F_1$ where $x \in E_1$. Since $B \cup x$ is connected, $B \cup x \subset E_1$, which implies that $F_1 \subset A$. If we let p_{n_2} be the first element in P which belongs to F_1, then there exists a separation $M \sim p_{n_2} = E_2 \cup F_2$ where $p_{n_1} \in E_2$. Since $E_1 \cup p_{n_1}$ is connected, we have $E_1 \cup p_{n_1} \subset E_2$, so that $F_2 \subset F_1$.

In general, for each $i > 1$, let p_{n_i} be the first element of P belonging to $F_{i-1} \subset F_{i-2} \subset \cdots \subset A$. Then there exists a separation $M \sim p_{n_i} = E_i \cup F_i$ where $p_{n_{i-1}} \in E_i$.

Since $F_1, F_2, \ldots$ is a monotone decreasing sequence satisfying $F_i \supset (F_{i+1} \cup p_{n_{i+1}}) = \mathrm{Cl}(F_{i+1})$, we have $F = \bigcap F_i \neq \emptyset$. If $E = \bigcup_1^\infty (E_i \cup p_{n_i})$, then F contains no point of E. Since $E_i \cup p_{n_i}$ is connected and contained in E_{i+1}, E is connected. Also, E contains P, since $E_{i+1} \supset \bigcup_1^{n_i} p_n$; hence $\mathrm{Cl}(E) = M$, and we have that each point of F is a limit point of E. Let $x \in F$; then $E \subset M \sim x \subset \mathrm{Cl}(E)$, implying that $M \sim x$ is connected. But then x is a noncut point, contrary to our assumption.

3. A nondegenerate metric continuum T is a simple arc from a to b if and only if every point of $T \sim a \sim b$ is a cut point of T.

The necessity is an immediate consequence of the properties of the unit interval. Now suppose that T is a metric continuum where each point of $T \sim a \sim b$ is a cut point of T, and let $x \in T \sim a \sim b$. Then there exists a separation $T \sim x = A \cup B$ where $a \in A$. Suppose $b \notin B$. Consider the continuum $B \cup x$. It must contain two noncut points; in particular it must contain a point $y \neq x$ such that $(B \cup x) \sim y$ is connected. Then $(A \cup x) \cup (B \cup x) \sim y = T \sim x$ is connected, and y is a noncut point. But $y \in B$, so that $y \neq a, b$, and we assumed that all points other than a and b were cut points. Hence $b \in B$, and by our previous characterization, T is a simple arc from a to b.

4. A nondegenerate metric continuum T is a simple closed curve if and only if it is disconnected by the removal of any two distinct points.

The necessity is clear. To prove the sufficiency we first note the following lemma:

Lemma. *A set is a simple closed curve if and only if it is the union of two simple arcs with the same end points and otherwise disjoint.*

Let T be a continuum where the removal of any two points causes T to be disconnected, and let p, q be two noncut points; i.e., $T \sim p$ and $T \sim q$ are connected. By our assumption we know that $T \sim p \sim q$ is not connected and there exists a separation $T \sim p \sim q = M \cup N$. It follows then that $p \cup M$, $p \cup N$, $q \cup M$, and $q \cup N$ are connected sets.

We claim that $\text{Cl}(M) \cup \text{Cl}(N) = T$. We know that $\text{Cl}(M \cup N) = \text{Cl}(M) \cup \text{Cl}(N) = \text{Cl}(T \sim p \sim q)$. Suppose p is not a limit point of $T \sim p \sim q$. Then there exists a separation $T \sim q = (T \sim p \sim q) \cup p$, which contradicts that $T \sim q$ is connected. Therefore p and, by symmetry, q are limit points of $T \sim p \sim q$, and $\text{Cl}(T \sim p \sim q) = T$.

We know that $\text{Cl}(M) \cap \text{Cl}(N) \neq \varnothing$, or T wouldn't be connected, and certainly $\text{Cl}(M) \cap \text{Cl}(N) \subset \{p,q\}$, since $M \cap N \neq \varnothing$. Suppose $\text{Cl}(M) \cap \text{Cl}(N) = \{q\}$ and $p \in \text{Cl}(M)$, $p \notin \text{Cl}(N)$. Considering the separation $T \sim q = (M \cup p) \cup N$, we have that $T \sim q$ is disconnected, contradicting the fact that q is a noncut point. Hence $\text{Cl}(M) \cap \text{Cl}(N) = \{p,q\}$.

Since $\text{Cl}(M) \subset T, \text{Cl}(M)$ is compact. We know that $\text{Cl}(M)$ is connected, since $\text{Cl}(M) = (M \cup p) \cup (M \cup q)$. Similarly, $\text{Cl}(N)$ is a continuum. We now show that $\text{Cl}(M)$ and $\text{Cl}(N)$ are simple arcs from p to q. Suppose neither of them is a simple arc from p to q. Then there exists a point $r \in \text{Cl}(M) \sim p \sim q$ such that $\text{Cl}(M) \sim r$ is connected, and a point $s \in \text{Cl}(N) \sim p \sim q$ such that $\text{Cl}(N) \sim s$ is connected. But then $T \sim r \sim s = (\text{Cl}(M) \sim r) \cup (\text{Cl}(N) \sim s)$ is connected, which is a contradiction. Therefore at least one of these sets, say $\text{Cl}(M)$, is a simple arc from p to q. Now suppose there exists a point $n \in \text{Cl}(N) \sim p \sim q$ such that $\text{Cl}(N) \sim n$ is connected. Let $m \in \text{Cl}(M) \sim p \sim q$. Then there exists a separation $\text{Cl}(M) \sim m = M_p \cup M_q$ where $p \in M_p, q \in M_q$, and we have that M_p and M_q are connected, since each is the continuous image of a half-open interval. Now we have that $M_p \cup M_q \cup (\text{Cl}(N) \sim n) = T \sim m \sim n$ is connected, but this is a contradiction. Hence $\text{Cl}(M)$ and $\text{Cl}(N)$ are both simple arcs from p to q, and by the lemma, $T = \text{Cl}(M) \cup \text{Cl}(N)$ is a simple closed curve.

SECTION XX

Arcwise Connectedness

Definition. A set M is said to be *arcwise connected* provided that if a, b is any point pair in M, then M contains a simple arc from a to b.

EXAMPLE 1. Consider the union of segments of unit length which in conjunction with the unit interval describe an infinite sequence of angles tending toward zero degrees, with vertex at the origin. This set is not locally connected but is arcwise connected. [See Example (iii), §XIV.]

EXAMPLE 2. Consider the set formed by taking the union of the set $\{(x,y): x = 0, \ -1 \le y \le 1\}$ and the set $\{(x,y): y = \sin 1/x, \ 0 < x \le 1\}$. This set is connected but is neither arcwise nor locally connected.

EXERCISES XX.

1. Every locally connected metric continuum is arcwise connected.

2. Every locally connected generalized metric continuum L is arcwise connected.

3. Every region in a locally connected metric continuum is arcwise connected.

SOLUTIONS.

1. Every locally connected metric continuum is arcwise connected.
  We first prove the following:

Lemma. *If a, b is any point pair in M, a locally connected metric continuum, then for any d > 0 we can join a to b with a simple chain $L_1, L_2, L_3, \ldots, L_N$ of locally connected continua of diameter < d such that $L_i \cdot L_j \neq \varnothing$ if and only if $|i - j| \leq 1$.*

PROOF. From Appendix I, we know that every point of M lies interior to a locally connected continuum of arbitrarily small diameter. Therefore, let us choose as a covering for M, the interior of a locally connected continua of diameter < d. By the compactness of M, we may take a finite subcovering, $\mathcal{G}$, of such sets. Now let R_1 be the continuum containing a where $\text{int}(R_1) \in \mathcal{G}$. Then choose a continuum $R_2 \neq R_1$ meeting R_1 with $\text{int}(R_2) \in \mathcal{G}$, and continue this process until a continuum R_N containing b has been incorporated, $\text{int}(R_N) \in \mathcal{G}$. This construction is possible; for if not, then M would be the countable union of two or more disjoint closed sets.

  Thus we have obtained a chain $R_1, R_2, \ldots, R_N$ of continua from a to b with successive links meeting. Now suppose R_K is the link of greatest subscript meeting R_1; then form the chain $R_1, R_K, R_{K+1}, \ldots, R_N$. Next, if R_M is the link of greatest subscript meeting R_K, form the chain $R_1, R_K, R_M, R_{M+1}, \ldots, R_N$. Repeating this process will eventually yield a chain from a to b whose successive links meet and nonsuccessive links do not meet.   □

  We now return to the proof of the proposition.
  Applying the lemma with $d = 1$, we obtain a finite chain $[R_1^{(1)}, R_2^{(1)}, \ldots, R_{N_1}^{(1)}]$ from a to b of locally connected continua of diameter less than one, such that $R_i^{(1)} \cdot R_j^{(1)} \neq \varnothing$ if and only if $|i - j| \leq 1$. Let $L_1 = \bigcup_{i=1}^{N_1} R_i^{(1)}$.
  Now, since $R_1^{(1)}$ is a locally connected continuum, we may get a chain with the above intersection properties from a to a point p_2 in $R_1^{(1)} \cdot R_2^{(1)}$ and such that only its last link meets $R_2^{(1)}$. We require that the diameter of these links be less than $\frac{1}{2}$, and that they be locally connected continua. In $R_2^{(1)}$, such a chain exists between p_2 in $R_1^{(1)} \cdot R_2^{(1)}$ and p_3 in $R_2^{(1)} \cdot R_3^{(1)}$, only the last link of which meets $R_3^{(1)}$. Continuing in this way, we obtain a simple chain $[R_1^{(2)}, R_2^{(2)}, \ldots, R_{N_2}^{(2)}]$ which has the properties of first chain with the additional stipulation that diam $R_i^{(2)} < \frac{1}{2}$. Also $L_2 = \bigcup_{i=1}^{N_2} R_i^{(2)}$ and $L_1 \supset L_2$, as L_2 is the union of N_1 simple chains $C_1^2, C_2^2, \ldots, C_{N_1}^2$, each contained in the corresponding link $R_n^{(1)}$ of the chain L_1.

  In general, for each n, we obtain $[R_1^{(n)}, \ldots, R_{N_n}^{(n)}]$ (which has the previously mentioned intersection property), diam $R_i^{(n)} < 1/n$, and $L_n = \bigcup_{i=1}^{N_n} R_i^{(n)} \subset L_{n-1}$.
  We now observe that for each n, L_n is a continuum, and then $L = \bigcap_{n=1}^{\infty} L_n$ is a nondegenerate continuum, which will be shown to be a simple arc from a to b. For suppose that $\alpha \in L \sim a \sim b$, and denote by $R_{\alpha_n}^n$ the first link in L_n containing α, and by $R_{\alpha_n'}^n$ the second and last link in L_n containing α. Observe that no three links of L_n can meet, so α is in at most two links. We agree to take $R_{\alpha_n}^n = R_{\alpha_n'}^n$ if α is in only one link. Then $\alpha = \bigcap_{n=1}^{\alpha} [R_{\alpha_n}^n + R_{\alpha_n'}^n]$, the intersection of monotone decreasing sequence of compact sets of diameter < 2/n about α. Define $H_n^{(a)} = \bigcup_{i=1}^{\alpha_n} R_i^{(n)}$ and $H_n^{(b)} = \bigcup_{j=\alpha_n}^{N_n} R_j^{(n)}$. We assert that $\cdots \supset H_n^{(a)} \supset H_{n+1}^{(a)} \supset \cdots$ and $\cdots \supset H_n^{(b)} \supset H_{n+1}^{(b)} \supset \cdots$. For assume that $R_K^{(n+1)}$, a link of $H_{n+1}^{(a)}$, includes a point of $L_n \sim H_n^{(a)}$. Then the subscript K is greater than that of all the $n + 1$st links $R_i^{(n+1)}$ in $[R_1^{(n)}, \ldots, R_{\alpha_n}^n, R_{\alpha_n'}^n]$. However, some $n + 1$st link in $[R_{\alpha_n}^n + R_{\alpha_n'}^n]$ must contain α, so $R_K^{(n+1)}$ cannot be in $H_{(n+1)}^{(a)}$, which was defined

to be the union of $n + 1$st links of subscript $\leq \alpha'_{(n+1)}$. Thus we have established that $\{H_n^{(a)}\}$ is a monotone decreasing sequence of sets; the same result similarly derived for $H_n^{(b)}$. Because each $H_n^{(a)}$ and $H_n^{(b)}$ is compact, we see that $H_a = \bigcap_{n=1}^{\infty} H_n^{(a)}$ and $H_b = \bigcap_{n=1}^{\infty} H_n^{(b)}$ are closed sets. Also $H_a + H_b = L$, as we see from

$$\bigcap_{n=1}^{\infty} H_n^{(a)} + \bigcap_{n=1}^{\infty} H_n^{(b)} = \bigcap_{n=1}^{\infty} [H_n^{(a)} + H_n^{(b)}]$$

$$= \bigcap_{n=1}^{\infty} L_n = L.$$

Likewise,

$$H_a \cdot H_b = \bigcap_{n=1}^{\infty} H_n^{(a)} \cdot \bigcap_{n=1}^{\infty} H_n^{(b)} = \bigcap_{n=1}^{\infty} [H_n^{(a)} \cdot H_n^{(b)}]$$

$$= \bigcap_{n=1}^{\infty} [R_{\alpha_n}^{(n)} + R_{\alpha_n'}^{(n)}] = \alpha.$$

Hence $L \sim \alpha = (H_a \sim \alpha) + (H_b \sim \alpha)$ is a separation since H_a and H_b are closed and meet only on the point α. Exercise 3 §XIX indicates that we have found the desired simple arc from a to b.

2. Every locally connected generalized metric continuum L is arcwise connected.

If a, b are points of L, then there exists a locally connected continuum M containing a and b. For if M_a denotes the set of points of L which lie together with a in a locally connected continuum, then M_a is open, since each point of L lies in an arbitrarily small locally connected region in L by Appendix I. However, M_a is also closed, for if p is a limit point of M_a, then p is interior to R_p, a locally connected continuum which then must meet M_a. Supposing that $q \in M_a \cdot R_p$, then q and a lie in a locally connected continuum K_q, so that $K_q \cup R_p$ is a locally connected continuum containing a and p. Then the preceding result applied to M yields an arc in L from a to b.

3. Every region in a locally connected generalized metric continuum is arcwise connected.

Such a region is itself a metric, locally connected, generalized continuum.

Localization of Property S

Definition. If M is a locally connected metric space, M *has property S locally* provided that for each point p in M, p belongs to an arbitrarily small region of M having property S itself.

Definition. For $r > 0$, a T_r-*chain in* M is a finite chain of connected sets $L_1, L_2, L_3, \ldots, L_n$ where $\delta(L_i) < r/2^i$ and successive links intersect. For $p \in M$, we define $T_r(p)$ to be the union of all T_r-chains from p. Evidently $\delta[T_r(p)] < 2r$.

Lemma. *If M is a metric space having property S, then so also does $T_r(p)$ have property S, for any $p \in M$.*

PROOF. Note that $T_r(p)$ exists for each p and each r. Let $\varepsilon > 0$ and $r > 0$ be given. Since M is locally connected, choose the least K such that $\sum_{n=K}^{\infty} r/2^N < \varepsilon/4$, and let $K(p)$ denote the union of all T_r-chains from p with no more than K links. Applying property S to M, obtain a finite cover of M by connected subsets of M with diameter $< r/2^{K+1}$, and let $W_1, W_2, \ldots, W_n$ be the sets of this cover which meet $K(p)$. If $W = \bigcup_{i=1}^{n} W_i$, then $K(p) \subset W \subset T_r(p)$, since W may be considered as a subset of the union of all T_r-chains from p with $< K + 1$ links.

Then for each i, let Q_i be the union of W_i together with all points in $T_r(p)$ which can be joined to W_i with a connected subset of $T_r(p)$ of diameter $< \varepsilon/4$. Clearly Q_i is a connected subset of $T_r(p)$ of diameter $< \varepsilon$. Thus $U = \bigcup_{i=1}^{n} Q_i$ is contained in $T_r(p)$. We claim that $T_r(p)$ is contained in U. The set $K(p)$ is clearly in U, since $K(p)$ is covered by $\{W_1, \ldots, W_n\}$, so we consider $x \in T_r(p) \sim K(p)$. Then $x \in L_m$ for $m > K$, where we have a chain L_1, $L_2, \ldots, L_K, L_{K+1}, \ldots, L_m$ with L_K meeting some W_i. Then the region

$L_{K+1} \cup \cdots \cup L_m$ has diameter $< \varepsilon/4$, so $x \in Q_i \subset U$. [The fact that $\delta(\bigcup_{i=K+1}^{m} L_i) < \varepsilon/4$ follows from our choice of K.] Hence $T_r(p) \subset U$, so $U = T_r(p)$, i.e., $T_r(p)$ can be expressed as the union of a finite number of connected sets of diameter $<\varepsilon$, and hence has property S. $\quad\square$

Theorem. *If M has property S, then M has property S locally.*

PROOF. Just note that $T_r(p)$ gives an arbitrarily small region having property S for any point of M and choice of $r > 0$, since $\delta(T_r(p)) < 2r$. $\quad\square$

Theorem. *If M is a locally connected locally compact metric space, then M has property S locally.*

PROOF. Let $p \in M$, and choose an arbitrarily small $\varepsilon > 0$ such that $\mathrm{Cl}(M \cdot V_\varepsilon(p))$ is compact. Then the proof that $T_\varepsilon(p)$ has property S follows from the proof of the lemma after we note that we can cover $\mathrm{Cl}(M \cdot V_\varepsilon(p))$ with a finite number of regions in M of diameter $<\varepsilon/2^{K+1}$. Thus $T_\varepsilon(p)$ is an arbitrarily small region of M about p having property S. $\quad\square$

Theorem. *If M is a locally connected, metric continuum, then each point of M lies interior to an arbitrarily small locally connected continuum.*

PROOF. Let $p \in M$ and $\varepsilon > 0$ be given. Then $T_\varepsilon(p)$ is connected, so $\mathrm{Cl}(T_\varepsilon(p))$ is a continuum of diameter $<\varepsilon$. $\mathrm{Cl}(T_\varepsilon(p))$ has property S, since $T_\varepsilon(p)$ does, and $\mathrm{Cl}(T_\varepsilon(p))$ is thus locally connected, and p is in the interior of $\mathrm{Cl}(T_\varepsilon(p))$. $\quad\square$

Cyclic Element Theory

Definitions. A point p is called a *cut point* (*local cut point*) of a connected space M if p separates M (separates an open set of M). A point p is called an *endpoint* of a connected space M if for every open set U of p there exists an open set V of p such that $p \in V \subset U$ and ∂V is a single point. With every noncut point q of M, we associate a set C_q which consists of all points $x \in M$ such that for no $y \in M$ is $M \sim y = M_1 \cup M_2$ a separation with $q \in M_1$ and $x \in M_2$. By a *cyclic element* of M will be meant either a cut point of M or a set C_q. A set C in a space M is said to be *cyclically connected* if every pair of points a and b in C lie on a simple closed curve J in C.

Our objective is to establish some basic properties of cyclic elements and to show that cyclic connectivity of a locally connected continuum M is equivalent to there being no cut point in M. For the lemmas and theorems in this appendix we assume M represents a locally connected continuum.

Lemma 1. *A point $q \in M$ is a C_q if and only if q is an end point.*

PROOF. If $C_q = q$, then q is not a cut point of M. Let U be an open set containing q. Since q is a noncut point, there exists an open set W of q such that $W \subset U$ and $M \sim W$ is connected. Let $[qx]$ be a simple arc such that $[qx] \cap (M \sim W) = x$. There is a $y \in M$ such that $M \sim y = M_1 \cup M_2$ is a separation with $q \in M_1$ and $x \in M_2$. Clearly $y \in [qx]$ and $y \notin M \sim W$, so that $M \sim W \subset M_2$ and $M_1 \subset W$. We take $M_1 = V$ and note $\partial V = y$. ☐

Whenever q is an end point, it follows from the definition that $C_q = q$.

Lemma 2. *Every set C_q is closed.*

PROOF. For $x \notin C_q$ there exists a y such that $M \sim y = M_1 \cup M_2$ is a separation, where $q \in M_1$ and $x \in M_2$. Since $C_q \subset M_1$ and M_2 is open, it follows that the complement of C_q is open.   □

Lemma 3. *For any C_q of M, each component R of $M \sim C_q$ has exactly one boundary point in C_q.*

PROOF. If $C_q = q$, then $R = M \sim q$ has q as its boundary point. Suppose C_q is nondegenerate and u and v are distinct boundary points of R in C_q. Let U and V be disjoint regions containing u and v respectively. In $R \cup U \cup V$ there exists a simple arc $[uv]$. The arc $[uv]$ has a subarc $[u'v']$ such that the open arc $(u'v')$ is in R and u' and v' are in C_q. Let w be a point of $[u'v']$ in R, and p a point such that $M \sim p = M_1 \cup M_2$ is a separation, where $q \in M_1$ and $w \in M_2$. Then $C_q \sim p \subset M_1$, and therefore u' or v' is in M_1. But then either $[u'w)$ or $[v'w)$ is in M_1, which is impossible.   □

Lemma 4. *If K is a connected subset of M, then $K \cap C_q$ is connected.*

PROOF. Suppose $K \cap C_q = A \cup B$ is a separation. Let $K(A)$ be the union of A and the points of K which are in a component of $M \sim C_q$ whose boundary point is in A. Define $K(B)$ similarly, and note that $K = K(A) \cup K(B)$, where $K(A) \cap K(B) = \varnothing$. Since K is connected, we can assume $K(A)$ has a limit point x in $K(B)$. The point x must be in B, as each point of $K(B) \sim B$ is in an open set which misses $K(A)$. Let W be a region about x which misses A. Then W meets $K(A) \sim A$, and consequently some component of $M \sim C_q$ has boundary points in A and in B, which contradicts Lemma 3.   □

Lemma 5. *Each C_q in M is a connected and locally connected subspace which has no cut point.*

PROOF. By Lemma 4, since M is connected, $M \cap C_q = C_q$ is connected. For $x \in C_q$ and W any region containing x, $W \cap C_q$ is, by Lemma 4, connected, which implies C_q is locally connected. Suppose there is an $x \in C_q$ such that $C_q \sim x = A \cup B$ is a separation. Let $a \in A$ and $b \in B$. The point q is a noncut point of M, so that $q \neq x$ by Lemma 4. The point x cannot separate a from b in M, for this would imply that x separates a or b from q in M, which is impossible. Thus if R is the component of $M \sim x$ containing a and b, then $R \cap C_q$ is a connected set joining a to b in C_q, which is a contradiction.   □

Theorem 1. *Every nondegenerate C_q is a locally connected continuum with no cut point.*

Lemma 6. *If A and B are disjoint nondegenerate closed sets in M, and M has no cut point, then there exist a pair of disjoint arcs from A to B.*

PROOF. Let $a' \in A$ and $b' \in B$ and let $[a'b']$ be an arc joining a' to b'. There exists an arc $[ab] \subset [a'b']$ such that $A \cap [ab] = a$ and $B \cap [ab] = b$. Let $p \in A \sim a$ and R a region containing p which misses $[ab]$. Then each $x \in R$ can be joined to p by an arc in R which then misses $[ab]$. Let K be the set of points $x \in M$ such that $x \in A$ or there exist disjoint arcs $[ab]$ and $[px]$ as above. We prove $K = M$ by showing that K is both open and closed in M. The set K is clearly open, since M is locally connected, so suppose y is a limit point of S. We can assume $y \notin A$, since $A \subset S$. Since y is a noncut point of M there exists in $M \sim y$ an arc $[cd]$ meeting A and B only in c and d respectively. Let W be a region containing y such that $\mathrm{Cl}(W) \cap (A \cup [c,d]) = \varnothing$, and let $x \in W \cap S$. Then there exist $[ab]$ and $[px]$ satisfying the condition above.

If $W \cap [ab] = \varnothing$, then $y \in S$. If $W \cap [ab] \neq \varnothing$, then $[ab]$ and $[px]$ contain arcs $[aw]$ and $[pr]$ respectively such that $[aw] \cap \mathrm{Cl}(W) = w$ and $[pr] \cap \mathrm{Cl}(W) = r$. Let $L = A \cup [aw] \cup [pr]$. Then $[cd]$ contains an arc $[uv]$, perhaps degenerate, such that $[uv] \cap L = u$ and $[uv] \cap B = v$. Let T denote one of the sets $[aw]$, $[pr]$ which does not contain u, and let H represent the other one of these sets. Let R be a region containing the point $T \cap \mathrm{Cl}(W)$ and containing no point of $H \cup [uv]$. Then $H \cup [uv]$ contains an arc $[st]$, and $T \cup R \cup W$ contains an arc $[qy]$, such that $[st] \cap A = s$, $[st] \cap B = t$, $[qy] \cap A = q$, and $[st] \cap [qy] = \varnothing$, so that $y \in S$. $\square$

Theorem 2 (Three-Point Theorem). *If a, p, and b are three distinct points of M, and M has no cut point, then there exists a simple arc $[apb]$ in M.*

PROOF. If p is a local cut point of M, then let $R \sim p = R_1 \cup R_2$ be a separation of a region R of p. If S_i is a component of R_i, $i = 1, 2$, then since $M \sim p$ is a locally connected generalized continuum, choose $d_i \in S_i$, $i = 1, 2$, and construct an arc $[d_1 d_2]$ in $M \sim p$. Then we can obtain an arc $[d_1 p]$ which meets the arc $d_2 p$ only at p, since S_1, S_2 are different components. Then $[d_1 p] \cup [pd_2] \cup [d_2 d_1]$ yields a simple closed curve J (after a possible replacement of d_i within S_i). Apply Lemma 6 to a, b, and J, and we get the desired arc $[apd]$.

If p is not a local cut point of M, we can suppose p separates no region in M. Since M has property S locally, we may choose E_1, an ε-region about p with property S and such that $a, b \notin \mathrm{Cl}(E_1)$. Since p is not a cut point of E_1, it is not a cut point of $\mathrm{Cl}(E_1)$. The point p is not an end point of $\mathrm{Cl}(E_1)$; otherwise it would be an end point of M. Thus the cyclic element C_p^1 of $\mathrm{Cl}(E_1)$ is nondegenerate. By Lemma 6 there exist in M two disjoint arcs $[aa_1]$ and $[bb_1]$ such that $[aa_1] \cap C_p^1 = a_1$ and $[bb_1] \cap C_p^1 = b_1$. Let E_2 be a region in C_p^1 with property S, of diameter less than $\varepsilon/2$, and such that $a_1, b_1 \notin \mathrm{Cl}(E_2)$. Once again we may suppose that p cuts no region in C_p^1; and it follows as in the case of $\mathrm{Cl}(E_1)$ that p lies in a nondegenerate cyclic element C_p^2 of $\mathrm{Cl}(E_1)$. By Lemma 6 there exist in C_p^1 two disjoint arcs $[a_1 a_2]$ and $[b_1 b_2]$ such that $[a_1 a_2] \cap C_p^2 = a_2$ and $[b_1 b_2] \cap C_p^2 = b_2$. Let E_3 be a region in C_p^2 with property S and of diameter less than $\varepsilon/3$, and so on. If we continue this process, the sets $A = \{\bigcup_{n=1}^{\infty} [a_n a_{n+1}] \cup \{p\}\}$ and $B = \{\bigcup_{n=1}^{\infty} [b_n b_{n+1}] \cup \{p\}\}$ are simple arcs meeting at p. (To verify that they are simple arcs from a_1 to p or b_1 to p, we observe that they are closed and hence compact, as p could be the only missing limit point of $\bigcup_{n=1}^{\infty} [a_n a_{n+1}]$ or $\bigcup_{n=1}^{\infty} [b_n b_{n+1}]$, and connected. Also they are each separated by the removal of any point other than a_1 (or b_1) and p.) Thus again we get an arc

$$[apb] = [aa_1] \cup \bigcup_{n=1}^{\infty} [a_n a_{n+1}] \cup \{p\}$$

$$\cup \bigcup_{n=1}^{\infty} [b_n b_{n+1}] \cup [b_1 b]. \qquad \square$$

Theorem 3. *A locally connected continuum M is cyclically connected if and only if it has no cut points.*

PROOF. The necessity is immediate, for if $M \sim p = A \cup B$ is a separation, then the points $a \in A$ and $b \in B$ are not on a simple closed curve in M.

Conversely, given a and b in M, where M has no cut point, then $a \in J$, a simple closed curve in M. This follows from the connectedness and hence arcwise connectedness of $M \sim p$, in the following way: Let $[pac]$ be an arc in M, and $[pqc]$ an arc in $M \sim a$. Then $[pac] \cup [pqc]$ contains a simple closed curve J containing a. In the event $b \notin J$, apply Theorem 2 to the points x, y, and b, where x and y are distinct points of J, to get an arc $[xby]$. Then a and b are contained in a simple closed curve in $J \cup [xby]$. $\qquad \square$

PART B

SECTION I

Product Spaces

We will define a particular topology on the cartesian space and develop relations between the topological properties of the product space and those of the factor spaces. In closing, we will give two examples of compactifications of a space.

The following definitions will be needed:

Definition. Given a collection of nonempty sets $[X_\mu]$ with $\mu \in M$, an index set, the *cartesian product set* $\prod_{\mu \in M} X_\mu$ is the set of selections $\{x_\mu\}$, where $x_\mu \in X_\mu$.

Definition. For each $\mu_0 \in M$, the function $p_{\mu_0} : \prod_{\mu \in M} X_\mu \to X_{\mu_0}$ by $p_{\mu_0}(\{x_\mu\}) = x_{\mu_0}$ is called the μ_0th *projection function.*

We can define a topology on $\prod_{\mu \in M} X_\mu$ if each X_μ is itself a topological space. Let the subbasic open sets be $p_\mu^{-1}(U_\mu)$, where U_μ is open in X_μ. Then take the basic sets in $\prod_{\mu \in M} X_\mu$ to be finite intersections of subbasic open sets. Then let the open sets in $\prod_{\mu \in M} X_\mu$ be generated by these basic sets. This is the topology we put on $\prod_{\mu \in M} X_\mu$.

The topology we have formulated corresponds to the norm topology in the case of finite product spaces. More definitions follow:

Definition. A *Tychonoff*, or *completely regular*, space X is a τ_1-space such that for each $x \in X$ and each closed set B not containing x, there exists a mapping $f : X \to [0,1]$ with $f(x) = 0$, $f(B) = 1$.

Definition. A family of mappings $f_\mu : X \to Y_\mu$ determines the function $H : X \to \prod_{\mu \in M} Y_\mu$ defined by $H(x) = \{f_\mu(x)\}$, $\mu \in M$.

Definition. A *compactification* of a topological space X is a pair (f,Y) where Y is a compact topological space and f is a homeomorphism of X onto a dense subset of Y.

Definition. The *one-point compactification* of a topological space X is the pair (i,X^*), where $X^* = X \cup \{\infty\}$ is the union of the set X and a single point $\{\infty\}$ not in X. The open sets of X^* are the open sets of X and the complements in X^* of closed, compact sets in X.

EXERCISES I.

1. The projection functions are continuous and open.

2. A sequence or directed family in a product space $X = \prod_{\mu \in M} X_\mu$ converges to a single point x if and only if for each μ the μth coordinate of the sequence or directed family converges to the μth coordinate of x.

3. A function $f : X \to Y = \prod_{\mu \in M} Y_\mu$ of a topological space into a product space is continuous if and only if either
   (a) f^{-1} of each subbasic open set is open;
   (b) $p_\mu \circ f : X \to Y_\mu$ is continuous for every $\mu \in M$.

4. A product space $X = \prod_{\mu \in M} X_\mu$ is Hausdorff if and only if each factor space X_μ is Hausdorff.

5. A product space $X = \prod_{\mu \in M} X_\mu$, where M is countable, is perfectly separable if and only if each factor space X_μ is perfectly separable.

6. A product space $X = \prod_{\mu \in M} X_\mu$ is connected if and only if each factor space X_μ is connected.

7. *Tychonoff Theorem*: A product space $X = \prod_{\mu \in M} X_\mu$ is compact if and only if each factor space X_μ is compact.

8. Given a family of mappings $f_\mu : X \to Y_\mu$ for each $\mu \in M$, define the function $H : X \to Y = \prod_{\mu \in M} Y_\mu$ by $H(x) = \{f_\mu(x)\}_{\mu \in M}$. Prove the following:
   (i) If each f_μ is continuous, then H is continuous.
   (ii) H is 1-1 provided that for $x_1 \neq x_2$ in X, there exists a mapping f_μ such that $f_\mu(x_1) \neq f_\mu(x_2)$ for some $\mu \in M$.
   (iii) If for each $a \in X$ and closed set B in $X \sim \{a\}$ there exists a $\mu \in M$ such that $f_\mu(a) \notin \text{Cl}(f_\mu(B))$, then H is an open mapping of X onto $H(X)$.

9. If X is a completely regular τ_1-space and $\mathcal{M} = [f_\mu]_{\mu \in M}$ is the family of all mappings of X into $[0,1]$, then $H(x) = \{f_\mu(x)\}_{\mu \in M}$, the mapping determined by $\mathcal{M}$, is a topological mapping of X into the cube $I^M = \prod_{\mu \in M} I_\mu$.

10. The pair $(H,\beta(X))$ is a maximal compactification of X, a completely regular space, where $\beta(X) = \text{Cl}(H(X))$, $H = \{f_\mu\}_{\mu \in M}$, and $\mathcal{M} = \{$all mappings of X into $[0,1]$ indexed by $M\}$.

11. The one-point compactification X^* of a topological space X is a compact topological space; X is dense in X^*; and X^* is Hausdorff if and only if X is Hausdorff and locally compact.

SOLUTIONS.

1. The projection functions are continuous and open.

Let $X = \prod_{\mu \in M} X_\mu$ be a product space with our established topology.

Let U be an open set in X_μ, an arbitrary factor space. Then $p_\mu^{-1}(U)$ is open in X by definition, so p_μ is continuous. Hence every projection function is continuous.

Next let A be open in X. Then $A = \bigcup_\alpha W_\alpha$, where each W_α is a basic set in X, so each W_α has the form $W_\alpha = V_{\alpha_1} \cap \cdots \cap V_{\alpha_n}$, where V_{α_i} is a subbasic open set in X, and each subbasic set is an inverse set relative to some X_μ. Therefore $p_{\mu_0}(A) = \bigcup_\alpha p_{\mu_0}(W_\alpha) = \bigcup_\alpha p_{\mu_0}(V_{\alpha_{\mu_0}})$. But $V_{\alpha_{\mu_0}} = p_{\mu_0}^{-1}(U_\alpha)$ for U_α an open inverse set in X_μ. Hence $p_{\mu_0}(A) = \bigcup_\alpha p_{\mu_0} \circ p_{\mu_0}^{-1}(U_\alpha) = \bigcup_\alpha U_\alpha$ or, if no $\mu = \mu_0$, $p_{\mu_0}(A) = X_{\mu_0}$. In either case, $p_{\mu_0}(A)$ is open. Therefore for any value of μ, p_μ is an open mapping.

2. A sequence or directed family in a product space $X = \prod_{\mu \in M} X_\mu$ converges to a single point $x \in X$ if and only if for each μ the μth coordinate of the sequence or directed family converges to the μth coordinate of x.

Let $\mathscr{D} = \{D_\alpha\}_{\alpha \in A}$ be a directed family in X, and suppose $\mathscr{D}$ converges to $x \in X$. Then for each μ, the function $p_\mu : X \to X_\mu$ is a mapping, so $p_\mu(\mathscr{D}) = \{(D_\mu)_\alpha\}$ converges to $p_\mu(x) = x_\mu$.

Conversely, suppose that for every $\mu \in M$, $\{(D_\mu)_\alpha\} \to x_\mu$. Let W be any basic open set about x. Then $W = V_{\mu_1} \cap \cdots \cap V_{\mu_n}$, where V_{μ_i} is an open subbasic set. Hence V_{μ_i} is an inverse set and $p_{\mu_i}(V_{\mu_i})$ is open in X_{μ_i}, so it must contain an element D_{μ_i} of the directed family $\{D_{\mu_i}\}_\alpha$ in X_{μ_i}. In particular, V_{μ_1} contains an element D_{α_1} of the directed family $\mathscr{D}$. Also D_{α_2} is an element of $\mathscr{D}$ in V_{μ_2}. However, $D_{\alpha_1} \cap D_{\alpha_2}$ contains some other element D_{α_k} of $\mathscr{D}$, so D_{α_k} is in $V_{\mu_1} \cap V_{\mu_2}$. By induction we get an element D_{α_0} of $\mathscr{D}$ in $V_{\mu_1} \cap \cdots \cap V_{\mu_n}$, since n is finite. Then the open basic set W contains an element of $\mathscr{D}$ namely D_{α_0}. Any open basic set about x contains such an element of $\mathscr{D}$, so $\mathscr{D}$ converges to x in X.

For sequences, just define a directed family $\mathscr{D} = \{D_n\}_{n \in I}$ where $D_n = (x_{n+1}, x_{n+2}, \ldots)$, the remainder after n terms of a sequence (x_n). Then apply the above to get the results.

3. A function $f : X \to Y = \prod_{\mu \in M} Y_\mu$ of a topological space into a product space is continuous if and only if either

(a) f^{-1} of each subbasic open set is open;
(b) $p_\mu \circ f : X \to Y_\mu$ is continuous for every $\mu \in M$.

Suppose $f : X \to \prod_{\mu \in M} Y_\mu$ is continuous. Then a subbasic open set U in Y is itself an open set in Y, so $f^{-1}(U)$ is open in X.

Conversely, suppose W is an open basic set in Y. It is enough to verify that $f^{-1}(W)$ is open in X. But $W = V_{\mu_1} \cap \cdots \cap V_{\mu_n}$ and $f^{-1}(W) = f^{-1}(V_{\mu_1}) \cap \cdots \cap f^{-1}(V_{\mu_n})$, a finite intersection of open sets, so it is open.

Suppose again that f is continuous. Then $p_\mu \circ f$ is a mapping of X to Y_μ, since p_μ is continuous for any $\mu \in M$.

Now let V_{μ_0} be an arbitrary open subbasic set in Y. Then $V_{\mu_0} = p_{\mu_0}^{-1}(U_{\mu_0})$, where U_{μ_0} is open in Y_{μ_0}. Since $p_{\mu_0} \circ f$ is continuous, then $f^{-1} \circ p_{\mu_0}^{-1}(U_{\mu_0}) = f^{-1}(V_{\mu_0})$ is open, so f^{-1} of any subbasic open set is open, and, by the above, f is continuous.

4. A product space $X = \prod_{\mu \in M} X_\mu$ is Hausdorff if and only if each factor space X_μ is Hausdorff.

If X is Hausdorff, each distinct pair of points can be separated by basic open sets. Assume that some factor space X_{μ_0} is not Hausdorff, and let p and q be distinct points of X_{μ_0} which cannot be separated by open sets in X_{μ_0}. Then the points x and y in X can be selected such that $x_\mu = y_\mu$, $\mu \neq \mu_0$, but $x_{\mu_0} = p$ and $y_{\mu_0} = q$. Hence x and y are distinct in X, so there must be disjoint open basic sets which separate them. But such disjoint open basic sets cannot involve subbasic sets from X_{μ_0}, since open sets in X_{μ_0} cannot separate p and q. Thus no open basic sets separate x and y, so X cannot be Hausdorff.

Now assume that each factor space X_μ is Hausdorff, and let x and y be any distinct pair of points in X. Then for some μ_k, $x_{\mu_k} \neq y_{\mu_k}$, so we get open sets U_{μ_k} and W_{μ_k} which separate x_{μ_k} and y_{μ_k} in X_{μ_k}. Then any basic open sets in X of the form $V_{\mu_1} \cap \cdots \cap p_{\mu_k}^{-1}(U_{\mu_k}) \cap \cdots \cap V_{\mu_n}$ and $V_{\mu_1} \cap \cdots \cap p_{\mu_k}^{-1}(W_{\mu_k}) \cap \cdots \cap V_{\mu_n}$ serve to separate x and y in X.

5. A product space $X = \prod_{\mu \in M} X_\mu$, where M is countable, is perfectly separable if and only if each factor space X_μ is perfectly separable.

Suppose X is perfectly separable. Then the basis $\{W_\alpha\}$ in X is countable. We claim that the collection $\{p_\mu(W_\alpha)\}$ is a countable basis for the open sets in X_μ. For, if U is any open set in X_μ, then $p_\mu^{-1}(U)$ is open in X and $p_\mu^{-1}(U) = \bigcup W_\alpha$ (a countable union). That is, $p_\mu^{-1}(U)$ can be regarded as the union of some subcollection of the basis for X. Since $p_\mu^{-1}(U)$ is an inverse set, $U = p_\mu(\bigcup W_\alpha) = \bigcup p_\mu(W_\alpha)$, so that U is the union of a subcollection of the collection $\{p_\mu(W_\alpha)\}$.

Conversely, suppose M is countable and for each μ, $\{B_\mu\}_\alpha$ ($\alpha \in A_\mu$) is a countable basis for the open sets of X_μ. Consider the countable collection $\mathcal{B} = \bigcup_{\mu \in M} \{p_\mu^{-1}(B_\mu)\}_{\alpha \in A_\mu}$, the union over M of the collections of inverses, $p_\mu^{-1}(B_\mu)$, of basic sets in X_μ. We will show that the countable collection $\beta = \{$all possible finite intersections of sets in $\mathcal{B}\}$ is a basis for X. Let U be open in X; then $U = \bigcup W_\delta$, where $W_\delta = V_{\mu_1} \cap \cdots \cap V_{\mu_n}$ and $V_{\mu_i} = p_{\mu_i}^{-1}(U_i)$, since V_{μ_i} is a subbasic set. Then U_i is open in X_{μ_i}, and X_{μ_i} is perfectly separable, so that $U_i = \bigcup_{\alpha \in I_i} (B_{\mu_i})_\alpha$; that is, U_i is the union of a subcollection of the basis for X_{μ_i}. Hence $V_{\mu_i} = \bigcup_{\alpha \in I_i} p_{\mu_i}^{-1}(B_{\mu_i})_\alpha$, so that we can express W_δ as

$$W_\delta = \left\{ \bigcup_{\alpha \in I_1} p_{\mu_1}^{-1}(B_{\mu_1})_\alpha \right\} \cap \cdots \cap \left\{ \bigcup_{\alpha \in I_n} p_{\mu_n}^{-1}(B_{\mu_n})_\alpha \right\}.$$

But this can be written as the union of all possible n-tuples of the sets $\{p_{\mu_1}^{-1}(B_{\mu_1})_\alpha\}, \ldots, \{p_{\mu_n}^{-1}(B_{\mu_n})_\alpha\}$, or

$$W_\delta = \bigcup_{\alpha \in I_1, \ldots, \alpha \in I_n} \{p_{\mu_1}^{-1}(B_{\mu_1})_\alpha \cap \cdots \cap p_{\mu_n}^{-1}(B_{\mu_n})_\alpha\}.$$

Since each W_δ is so expressible, the set $U = \bigcup W_\delta$ is itself the union of a subcollection of the class β. Thus β is a countable basis for X.

6. A product space $X = \prod_{\mu \in M} X_\mu$ is connected if and only if each factor space X_μ is connected.

The necessity of the condition is immediate, since the projection functions are continuous. In order to simplify the proof of the sufficiency, we first impose an order on the set M. We will show that any two points a and b in X lie in a connected set. For any fixed μ_k in M, define the set $D_{\mu_k} = \{x \in X : x_\mu = a_\mu$ for $\mu < \mu_k$, $x_\mu = b_\mu$ for $\mu > \mu_k\}$. We claim that D_{μ_k} is connected; for if not, suppose A_k is a proper subset of D_{μ_k} which is open and closed in D_{μ_k}. Then $B_k = D_{\mu_k} - A_k$ is open and closed in D_{μ_k}, so $p_{\mu_k}(A_k)$ and $p_{\mu_k}(B_k)$ are each open in X_{μ_k}. However, $p_{\mu_k}(A_k) \cup p_{\mu_k}(B_k) = X_{\mu_k}$, a connected set which cannot be the union of two disjoint open sets. We know that $p_{\mu_k}(A_k)$ and $p_{\mu_k}(B_k)$ are disjoint, since A_k and B_k are disjoint and are inverse sets relative to p_{μ_k}. Therefore D_{μ_k} is connected for all k.

The set $\bigcup_{\mu \in M} D_\mu$ is then connected, since any two D_μ's intersect. Then we claim that a and b are contained in $\mathrm{Cl}(\bigcup_{\mu \in M} D_\mu)$. If M is finite, then $a \in D_{\mu_0}$, where $\mu_0 = \max\{\mu : \mu \in M\}$. If M is infinite and W is any open basic set which contains a, where $W = V_{\mu_1} \cap \cdots \cap V_{\mu_n}$, then $D_\mu \subset W$ for $\mu > \max\{\mu_1, \mu_2, \ldots, \mu_n\}$. Therefore W meets $\bigcup_{\mu \in M} D_\mu$, so $a \in \mathrm{Cl}(\bigcup_{\mu \in M} D_\mu)$. Also $b \in \mathrm{Cl}(\bigcup_{\mu \in M} D_\mu)$ in the same way, so that a and b are contained in the connected set $\mathrm{Cl}(\bigcup_{\mu \in M} D_\mu)$.

The next exercise, the Tychonoff theorem on compact product spaces, requires the introduction of some new concepts and the use of the Hausdorff Maximality Principle.

Definition. A binary relation $\le$ is said to be a *partial ordering* on a set P provided that for any elements x, y, and z in P, we have

(1) $x \le x$;
(2) $x \le y$ and $y \le x$ imply that $x = y$;
(3) $x \le y$ and $y \le z$ imply that $x \le z$.

Definition. A set P together with the partial ordering $\le$ is said to be a *partially ordered set*.

Definition. A subset C of a partially ordered set P is *simply ordered chain* provided, that for every x, y in C, $x \le y$ or $y \le x$.

Hausdorff Maximality Principle. *If P is a partially ordered set, then every $x \in P$ lies in a simply ordered chain C in P, and every simply ordered chain C in P is contained in a maximal simply ordered chain.*

EXAMPLE (Hausdorff Maximality Principle). If P is the collection of closed line segments in the unit disk, then there is no longest such line, but every simply ordered chain is contained in a maximal simply ordered chain.

Definition. A class of subsets of a space X is *deficient* if it fails to cover X, and *finitely deficient* if no finite subcollection covers X.

Remark. A set X is compact if and only if every finitely deficient collection of open subsets of X is also deficient.

Lemma 1. *Let $\mathscr{F}$ be a finitely deficient collection of open sets in X. There exists a maximal finitely deficient family of open sets in X containing $\mathscr{F}$.*

PROOF. Define $\mathscr{C}$ to be the collection of all finitely deficient families of open sets of X and partially order $\mathscr{C}$ by inclusion. Let $\mathscr{A}$ be a maximal simply ordered chain in $\mathscr{C}$ containing $\mathscr{F}$. Define $\mathscr{D} = \bigcup_{A \in \mathscr{A}} A$, and observe that $\mathscr{D}$ is finitely deficient. For if not, assume the sets $A_1, A_2, \ldots, A_n$ are an open cover for X. Then each A_i is an element in some C_i in $\mathscr{A}$, and $\mathscr{A}$ is simply ordered by set inclusion. Hence C_j contains $C_1, \ldots, C_n$

for some j, $1 \leq j \leq n$. Thus $A_1, \ldots, A_n$ is a finite subcollection of C_j, so $A_1, \ldots, A_n$ cannot cover X, since C_j is finitely deficient.

Also $\mathscr{D}$ is maximal, for if $\mathscr{D} + \{G\}$ is finitely deficient where G is an open set in X not included in $\mathscr{D}$, then $\mathscr{A}^* = \mathscr{A} + \{\mathscr{D} + (G)\}$ is a simply ordered chain containing $\mathscr{F}$, which contradicts the maximality of $\mathscr{A}$. $\square$

Remark. A maximal deficient or finitely deficient family of open sets must contain all of the open subsets of each of its sets.

Lemma 2. *If $\mathscr{D}$ is a maximal finitely deficient family of open subsets of X, and $G_1 \cap \cdots \cap G_m$ is an intersection of open sets in X such that $G_1 \cap \cdots \cap G_m$ lies in some $G \in \mathscr{D}$, then for some k $(1 \leq k \leq m)$ we have $G_k \in \mathscr{D}$.*

PROOF. Suppose the assertion is false, that is, no G_k is in $\mathscr{D}$. Note that if G_k is added to $\mathscr{D}$, the resulting family is not finitely deficient. Then it must be true that for each k $(1 \leq k \leq m)$, $\bigcup_{i \in I_k} A_i + G_k$ is a finite open cover for X, where I_k is some finite index set, $A_i \in \mathscr{D}$. However, this implies that the set $\bigcup_{i \in I_0} A_i + (G_1 \cap \cdots \cap G_m)$ is a finite open cover for X, where $I_0 = I_1 \cup \cdots \cup I_m$. Since each A_i is in $\mathscr{D}$ and $G_1 \cap \cdots \cap G_m$ is in some G in $\mathscr{D}$, then $\bigcup_{i \in I_0} A_i + G$ is a finite open cover for X. This contradicts the assumption that $\mathscr{D}$ is finitely deficient. Therefore some G_k must lie in G in $\mathscr{D}$, and the lemma is proved. $\square$

Lemma 3. *If every finitely deficient family of sub-basic open subsets of X is deficient, then X is compact.*

PROOF. Let $\mathscr{F}$ be any finitely deficient open covering of X, and let $\mathscr{D}$ be a maximal finitely deficient open covering of X which contains $\mathscr{F}$. Let $\mathscr{G}$ be the collection of subbasic open sets of X. Consider the collection $\mathscr{M} = \mathscr{D} \cap \mathscr{G}$. $\mathscr{M}$ is clearly finitely deficient, since $\mathscr{D}$ is. By assumption $\mathscr{M}$ cannot cover X, so we may choose some point $x_0 \in X$ that is not covered by $\mathscr{M}$. Since $\mathscr{D}$ is a cover, there is some open set $G \in \mathscr{D}$ such that $x_0 \in G$; then for some finite collection $S_1, \ldots, S_n$ of subbasic open sets, we have $x_0 \in \bigcap_{i=1}^{n} S_i \subset G$. By the preceding lemma, some S_j must be an element of $\mathscr{D}$ and hence an element of $\mathscr{M}$. Evidently this is a contradiction, showing that the alleged finitely deficient open cover $\mathscr{F}$ with which we began cannot exist. Thus X is compact. $\square$

7. *Tychonoff Theorem*: A product space $X = \prod_{\mu \in M} X_\mu$ is compact if and only if each factor space X_μ is compact.

Let $\mathscr{H}$ be any finitely deficient family of open subbasic sets covering X. Then for each $\mu \in M$, the set $\mathscr{H}_\mu = \{V_\mu : V_\mu$ is an open set in X_μ and $p_\mu^{-1}(V_\mu) \in \mathscr{H}\}$ is a finitely deficient open cover for X_μ, and hence deficient, since X_μ is compact. Therefore for each μ, choose x'_μ not in any element of $\mathscr{H}_\mu$. Then the point $x = (x'_\mu)$ is not in any set of $\mathscr{H}$, so $\mathscr{H}$ is deficient and X is compact.

Conversely, it is easy to see that $X_\mu = p_\mu(X)$ is compact for each $\mu \in M$, since p_μ is a mapping.

Remark. The projection functions are not closed, as is shown by the following example. In the product space $E^1 \times E^1 = E^2$, consider the closed set formed by taking a branch of a hyperbola, say $xy = 1$ in the first quadrant, and all the points "above" this branch. Call this set C; C is closed because it contains all its limit points, but its projection $p_x(C)$ in E^1 is the open set $(0, \infty)$.

8. Given a family of mappings $f_\mu : X \to Y_\mu$ for each $\mu \in M$, define the function $H : X \to Y = \prod_{\mu \in M} Y_\mu$ by $H(x) = \{f_\mu(x)\}_{\mu \in M}$. Prove the following:

(i) If each f_μ is continuous, then H is continuous.
(ii) H is 1-1 provided that for $x_1 \neq x_2$ in X, there exists a mapping f_μ such that $f_\mu(x_1) \neq f_\mu(x_2)$ for some $\mu \in M$.
(iii) If for each $a \in X$ and closed set B in $X \sim \{a\}$ there exists a $\mu \in M$ such that $f_\mu(a) \notin \mathrm{Cl}(f_\mu(B))$, then H is an open mapping of X onto $H(X)$.

The proof of (i) follows directly from Exercise 3(b) of this section, since $p_\mu \circ H = f_\mu$.

For (ii), evidently if $x_1 \neq x_2$ but $f_\mu(x_1) = f_\mu(x_2)$ for every μ, then $H(x_1) = H(x_2)$, so that H is not 1-1. Conversely, if $x_1 \neq x_2$ implies that $f_\mu(x_1) \neq f_\mu(x_2)$ for some μ, then $\{f_\mu(x_1)\}_{\mu \in M} \neq \{f_\mu(x_2)\}_{\mu \in M}$, so that H is 1-1.

To prove (iii) we first note that if B is a subset of X, then $\mathrm{Cl}(H(B)) \subset \prod_{\mu \in M} \mathrm{Cl}(f_\mu(B))$. This is easily verified: Let p be a limit point of $H(B)$, and form the directed family $\mathscr{F} = \{H(B) \cap U : U$ is an open set about $p\}$. Since $\mathscr{F} \to p$ in Y, we have that $p_\mu(\mathscr{F}) \to p_\mu$ in Y_μ. But this implies that p_μ is a limit point of $f_\mu(B)$ or $p_\mu \in f_\mu(B)$, so $p_\mu \in \mathrm{Cl}(f_\mu(B))$.

Now let U be any open set in X and $a \in U$. Then $a \notin X \sim U$, a closed set, so for some μ, $f_\mu(a) \notin \mathrm{Cl}(f_\mu(X \sim U))$. Hence $H(a) \notin \prod_{\mu \in M} \mathrm{Cl}(f_\mu(X \sim U))$.

Therefore, since $a \in U$ was arbitrary, $H(U) \subset Y \sim \mathrm{Cl}(H(X \sim U))$. However, the inclusion sign may be reversed, since $H(U) \supset [Y \cap H(X)] \sim [\mathrm{Cl}(H(X \sim U)) \cap H(X)]$. Therefore, $H(U) = [Y \cap H(X)] \sim [\mathrm{Cl}(H(X \sim U)) \cap H(X)]$, so that H is an open mapping of X onto $H(X)$.

Remark. A topological mapping of X into Y is a mapping h such that $h : X \to Y$ is a homeomorphism of X onto $h(X) \subset Y$. We use the notation $h : X \Rightarrow Y$ to represent an onto function.

9. If X is a completely regular τ_1-space and $\mathscr{M} = [f_\mu]_{\mu \in M}$ is the family of mappings of X into $I = [0,1]$, then $H(x) = \{f_\mu(x)\}_{\mu \in M}$, the mapping determined by $\mathscr{M}$, is a topological mapping of X into the cube $I^M = \prod_{\mu \in M} I_\mu$.

The function $H : X \to H(X)$ is onto and continuous by the preceding theorem, since each f_μ is continuous. Also H is 1-1, for let $x_1 \neq x_2$ be points in X. Then since X is τ_1, x_1 is contained in an open set A which does not contain x_2. Then $x_2 \in X - A$, a closed set, so since X is completely regular, we have a mapping f_μ such that $f_\mu(x_1) = 0$ and $f_\mu(X \sim A) = 1$, so $f_\mu(x_1) \neq f_\mu(x_2)$. Thus the preceding theorem indicates that H is 1-1. Finally H is an open mapping, for since X is completely regular, if $a \in X$ and $a \notin B$, a closed set, then for some μ, $f_\mu(a) = 0$, $f_\mu(B) = 1$; so clearly $f_\mu(a) \notin \mathrm{Cl}(f_\mu(B))$. By (iii) of the preceding theorem H is open. Then we have established that $H : X \to H(X)$ is a homeomorphism, since H is 1-1, onto, continuous, and open.

10. The pair $(H, \beta(X))$ is a maximal compactification of X, a completely regular space, where $\beta(X) = \mathrm{Cl}(H(X))$, $H = \{f_\mu\}_{\mu \in M}$, and $\mathscr{M} = \{$all mappings of X into $[0,1]$ indexed by $M\}$.

We know that $\beta(X)$ is compact, since $\mathrm{Cl}(H(X)) \subset \prod_{\mu \in M} \mathrm{Cl}(f_\mu(X))$, as we verified in a previous exercise, and $\prod_{\mu \in M} \mathrm{Cl}(f_\mu(X))$ is compact since each $\mathrm{Cl}(f_\mu(X))$ is compact, being a closed and bounded set of real numbers. In the preceding exercise we verified that $H : X \to H(X)$ is a homeomorphism, and since $H(X)$ is dense in $\beta(X)$, then $(H, \beta(X))$ is a compactification.

Let (h, Y) be a Hausdorff compactification of X. Then there exists a mapping $G : \prod_{\mu \in M} I \to \prod_{\alpha \in M_1} I$, where $\mathscr{M}_1 = \{$all mappings of Y into $[0,1]$ indexed by $M_1\}$, such that the following diagram is commutative:

$$\begin{array}{ccc} X & \xrightarrow{h} & Y \\ {\scriptstyle H}\downarrow & & \downarrow{\scriptstyle H_1} \\ \displaystyle\prod_{\mu\in M} I\mu & \xrightarrow{G} & \displaystyle\prod_{\alpha\in M_1} I\alpha \end{array}$$

For $f_\alpha : Y \to [0,1]$, we have $f_\alpha \circ h = f_\nu$ for some $\nu \in M$, so define $g_\alpha : \prod_{\mu\in M} I_\mu \to I_\alpha$ by $g_\alpha(\{t_\mu\}) = t_\nu \in I_\alpha$. The mapping $G = \{g_\alpha\}$ is a continuous mapping of $\prod_{\mu\in M} I_\mu$ into $\prod_{\alpha\in M_1} I_\alpha$ with the property that $G(\text{Cl}(H(X))) = H_1(Y)$ and $G \circ H$ is a homeomorphism of X onto $(H_1 \circ h)(X)$, which is dense in $H_1(Y)$.

11. The one-point compactification X^* of a topological space X is a compact topological space; X is dense in X^*; and X^* is Hausdorff if and only if X is Hausdorff and locally compact.

We designate open sets of X^* which are open in X as sets of type α, and open sets of X^* which are complements of closed, compact sets in X as sets of type β. To show that X^* is a topological space, we must have unions of open sets open and the intersection of a finite number of open sets open. The union of sets of type α is clearly open, and the union of sets of type β is open because it is the complement of a compact set of X. Let U_α, U_β be open sets of each kind. Then $U_\alpha \cup U_\beta$ is open, since $X^* - (U_\alpha \cup U_\beta) = (X^* - U_\alpha) \cap (X^* - U_\beta)$ is a compact set in X. Similarly, the intersection of a finite number of sets of type α or type β is open, and $U_\alpha \cap U_\beta$ is open because $U_\beta \cap X$ is open, being the complement of a closed set, and U_α is assumed to be open in X.

To show that X^* is compact, we first note that in any open cover of X^*, some open set G_0 must contain the point ∞, and the complement of this set is compact in X. Let $\{G_\alpha\}$ be an open cover of X^*. The sets $G_\alpha \cap X$ are open in X and cover $X \sim G_0$, a compact set. Hence we obtain a finite subcover $\{G_{\alpha_i} \cap X\}_{i=1,\ldots,N}$ for $X \sim G_0$, and $\{G_{\alpha_i}\} \cup \{G_0\}$ covers X^*.

That X is dense in X^* follows from the fact that every open set in X^* about ∞ meets X.

Suppose X is Hausdorff and locally compact. Let p be a point of X and U be an open set about p whose closure is compact. Then the open sets U and $X^* \sim \text{Cl}(U)$ separate p and ∞, so X^* is Hausdorff.

Conversely, if X^* is Hausdorff, clearly X is, since the Hausdorff property is hereditary (i.e. every subspace is Hausdorff). Now if $p \in X$, let the sets U and V be a Hausdorff separation of p and ∞ respectively. Then $X^* \sim V$ is a compact set in X about p, or X is locally compact.

Decomposition Spaces

We say that a topological space X *decomposes* into the collection $\mathscr{G}$ of subsets of X provided that $X = \bigcup_{g \in \mathscr{G}} g$ and these sets g are nonempty and disjoint.

Definition. The *natural function* φ of the decomposition $\mathscr{G}$ of X is the function $\varphi : X \to \mathscr{G}$ defined by the action $\varphi(x) = g_x$, where g_x is the element of $\mathscr{G}$ containing x.

Definition. The *natural decomposition* in X of a mapping $f : X \to Y$ is the collection of disjoint sets $\mathscr{G} = \{ f^{-1}(y) \}$, $y \in Y$.

Definition. Let Q be the set $\mathscr{G}$ considered as a topological space, with U open in Q if and only if $\varphi^{-1}(U)$ is open in X. The topological space Q is called the *quotient space* of $\mathscr{G}$.

Remarks.

(i) Q has the largest topology for which the natural function $\varphi : X \to Q$ is continuous.

(ii) The natural mapping $\varphi : X \to Q$ of any decomposition is quasicompact and onto.

Definition. *A decomposition $\mathscr{G}$ is said to be upper semicontinuous provided that the union of all elements of $\mathscr{G}$ contained in any open set of X is open. A decomposition $\mathscr{G}$ of X is lower semicontinuous provided that the union of all elements of $\mathscr{G}$ meeting any open sets is itself open.*

Definition. *A function $f : X \to Y$ is said to be* monotone *provided that the* set $f^{-1}(y)$ is a continuum in X for $y \in Y$.

93

EXERCISES II.

1. The quotient space Q of $\mathscr{G}$ is a τ_1-space if and only if the elements of $\mathscr{G}$ are closed sets.

2. The quotient space Q of a decomposition $\mathscr{G}$ of closed sets may fail to be Hausdorff even though X is Hausdorff and compact.

3. (a) Given $X \times Y$, $\mathscr{G} = \{p_X^{-1}(x):x \in X\}$, show that $Q \cong X$.
   (b) Given $X \times Y \times Z$, $\mathscr{G} = \{p_X^{-1}(x) \cap p_Y^{-1}(y):x \in X, y \in Y\}$, show that $Q \cong X \times Y$.

4. Given a mapping $f: X \Rightarrow Y$ where X and Y are τ_1-spaces, let $\mathscr{G}$ be the natural decomposition of f, Q the quotient space of $\mathscr{G}$, and φ the natural mapping from X to Q. The function $h:Q \to Y$ defined by $h = f \circ \varphi^{-1}$ is 1-1 and continuous. Further, h is a homeomorphism if and only if f is quasicompact.

5. Any upper-semicontinuous decomposition $\mathscr{G}$ of a τ_1 space has closed elements. Thus Q is a τ_1-space.

6. (a) If $f: X \to Y$ is quasicompact and the decomposition $\mathscr{G}$ of f is lower semi-continuous, then f is an open mapping.
   (b) If $f: X \Rightarrow Y$ is quasicompact and onto, and the natural decomposition $\mathscr{G}$ of f is upper semicontinuous, then f is a closed mapping.
   (c) Let Q be the quotient space of the natural decomposition $\mathscr{G}$ of a τ_1-space X.
      (1) Q is Hausdorff if X is normal and $\mathscr{G}$ is upper semicontinuous.
      (2) Q is normal if X is normal and $\mathscr{G}$ is upper semicontinuous.
      (3) Q is perfectly separable if X is perfectly separable and $\mathscr{G}$ is lower semi-continuous.
      (4) Q is metric if X is separable and metric and $\mathscr{G}$ is both upper and lower semicontinuous.

7. The quotient space Q of a decomposition $\mathscr{G}$ of a compact Hausdorff space X is itself Hausdorff if and only if $\mathscr{G}$ is upper semicontinuous.

8. Let X be Hausdorff and $f: X \Rightarrow Y$ a quasi-compact mapping with compact point-inverse frontiers. If the natural decomposition of f is upper semicontinuous, then Y is Hausdorff.

9. Given a monotone mapping $f: X \Rightarrow Y$ where X and Y are Hausdorff and X is locally compact, then the natural decomposition of f is upper semicontinuous.

10. If $f: X \Rightarrow Y$ is a quasicompact monotone mapping where X is locally compact and Hausdorff, then Y is Hausdorff if and only if the natural decomposition $\mathscr{G}$ of f is upper semicontinuous.

11. If the decomposition $\mathscr{G}$ of a space X is upper semicontinuous and has compact element frontiers, and if X is separable and metric, so also is the quotient space Q of $\mathscr{G}$.

SOLUTIONS.

1. The quotient space Q of $\mathscr{G}$ is a τ_1-space if and only if the elements of $\mathscr{G}$ are closed sets.

Suppose first that Q is a τ_1-space. Let g_0 be a fixed element of $\mathscr{G}$, and g any other element of $\mathscr{G}$. Then g is in an open set U of Q which does not meet g_0. The union of all such sets U is an open set, $\bigcup_{g \neq g_0} U_g = Q \sim \{g_0\}$. Therefore $\varphi^{-1}(Q \sim \{g_0\})$ is open in X, so g_0 is closed in X.

Conversely, if every element of $\mathscr{G}$ is closed in X, then if g_1, g_2 are distinct elements, we have $X \sim g_1$ an open set not meeting g_1 but containing g_2, and $X \sim g_1$ is an inverse set relative to φ, so that $\varphi(X \sim g_1) = Q \sim g_1$ is open in Q and contains g_2. Similarly, we can get an open set about g_1 which doesn't contain g_2.

2. The quotient space Q of a decomposition space $\mathscr{G}$ of closed sets may fail to be Hausdorff even though X is Hausdorff and compact.

Let X be the unit square $I \times I$ and decompose X accordingly: the line $x = r$, r rational, is a single element in $\mathscr{G}$; each point on the line $x = i$, i irrational, is an element in $\mathscr{G}$. We cannot get a Hausdorff separation of two points on a line with irrational x-coordinate in Q, since the inverse images of these open sets will meet a line with rational x-coordinate in X, and thus their images under φ will intersect in the decomposition. Then $\mathscr{G}$ is not upper semicontinuous, since if so, then some open sets in Q would have to miss every line (regarded as a single element). But this is impossible.

3. (a) Given $X \times Y$, $\mathscr{G} = \{p_X^{-1}(x) : x \in X\}$, show that $Q \cong X$.

(b) Given $X \times Y \times Z$, $\mathscr{G} = \{p_X^{-1}(x) \cap p_Y^{-1}(y) : x \in X, y \in Y\}$, show that $Q \cong X \times Y$.

To prove part (a), first note that the elements of $\mathscr{G}$ are of the form $x \times Y$. We want a homeomorphism f of Q onto X.

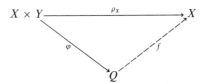

Define $f = p_X \circ \varphi^{-1}$. The elements of Q are of the form $x \times Y$, so it is easy to verify that $f : Q \to X$ is 1-1, well defined, and onto. Also f is open, for if $U \times Y$ is open in Q, then $p_X \circ \varphi^{-1}(U \times Y) = p_X(U \times Y)$, where U is open in X, so $p_X(U \times Y) = U$ is open in X. Also, f is continuous, since if V is open in X, then $f^{-1}(V) = \varphi \circ p_X^{-1}(V) = \varphi(V \times Y)$, and $V \times Y$ is an inverse set relative to φ, so $\varphi(V \times Y)$ is open in Q. Thus f is a homeomorphism.

The elements of $\mathscr{G}$ in part (b) are lines which extend over Z and hit the point (x, y) in the $X \times Y$ plane of the space $X \times Y \times Z$. We need a homeomorphism of Q onto $X \times Y$. Define $f = (p_X \times p_Y) \circ \varphi^{-1}$:

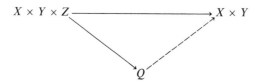

We point out that elements in Q have the form $x \times y \times Z$ for $(x, y) \in X \times Y$. It is left as an exercise to verify that f is 1-1, well defined, onto, continuous, and open, and therefore is a homeomorphism.

4. Given a mapping $f: X \Rightarrow Y$ where X and Y are τ_1-spaces, let $\mathscr{G}$ be the natural decomposition of f, Q the quotient space of $\mathscr{G}$, and φ the natural mapping from X to Q. The function $h: Q \to Y$ defined by $h = f \circ \varphi^{-1}$ is 1-1 and continuous. Also h is a homeomorphism if and only if f is quasicompact.

We know that h is well defined, for any element g of Q has, as its inverse under φ, the equivalence class of points in X which map into a single point under f. Clearly h is onto, since φ^{-1} and f are onto, and h is 1-1, since distinct elements in Q are mapped by φ^{-1} into disjoint equivalence classes in X relative to f. Now h is continuous, for if V is any open set in Y, then $h^{-1}(V) = \varphi \circ f^{-1}(V)$ and $f^{-1}(V)$ is open, since f is a mapping. Also $f^{-1}(V)$ is an inverse set relative to φ, since f and φ have the same inverse sets in X. But φ is quasicompact, so $\varphi(f^{-1}(V))$ is open in Q.

Furthermore, if f is quasicompact, then h is an open mapping, since $\varphi^{-1}(U)$ is an inverse set relative to f in X, for any set U in Q. In particular, if U is open in Q, $h(U) = f \circ \varphi^{-1}(U)$ is open in Y. But if h is an open mapping, then it is a homeomorphism. Now, suppose h is a homeomorphism. Then it is an open mapping, so that $h(U)$ is open in Y for any set U open in Q; thus $f(\varphi^{-1}(U))$ is open, so f maps open inverse sets into open sets and hence is quasicompact.

5. Any upper-semicontinuous decomposition $\mathscr{G}$ of a τ_1-space has closed elements, and thus Q is a τ_1-space.

If the topological space X is a τ_1-space and p is an element of X, then $X \sim p$ is open, so $X \sim g_p$ is open if g_p is the element of $\mathscr{G}$ containing p. Hence g_p is a closed inverse set in X, so $\varphi(g_p)$ is closed in Q, and Q is τ_1.

6. (a) If $f: X \to Y$ is quasicompact and the natural decomposition determined by f is lower semicontinuous, then f is an open mapping.
  (b) If $f: X \Rightarrow Y$ is quasicompact and onto, and the natural decomposition $\mathscr{G}$ determined by f is upper semicontinuous, then f is a closed mapping.
  (c) Let Q be the quotient space of the natural decomposition $\mathscr{G}$ of a τ_1-space X.
    (1) Q is Hausdorff if X is normal and $\mathscr{G}$ is upper semicontinuous.
    (2) Q is normal if X is normal and $\mathscr{G}$ is upper semicontinuous.
    (3) Q is perfectly separable if X is perfectly separable and $\mathscr{G}$ is lower semicontinuous.
    (4) Q is metric if X is separable and metric and $\mathscr{G}$ is both upper and lower semicontinuous.

For the proof of part (a), let U be an open set in X. Then $G = \bigcup g$ (where $g \in \mathscr{G}$ and g meets U) is an open set, since $\mathscr{G}$ is lower semicontinuous. Clearly G is an inverse set and $f(G) = f(U)$. Thus $f(G)$ being open in Y implies that f is an open mapping.

For part (b), let K be a closed set in X. Then $X \sim K$ is open, and if $G = \bigcup g$, where $g \in \mathscr{G}$ and $g \subset X \sim K$, then G is open by the upper semicontinuity of $\mathscr{G}$. Thus $X \sim G$ is a closed inverse set and $f(K) = f(X \sim G)$, so f is a closed mapping.

To show that Q is Hausdorff, let g_1, g_2 be distinct points of Q. Then $\varphi^{-1}(g_1)$ and $\varphi^{-1}(g_2)$ are disjoint closed sets in X, since Q is τ_1. (This follows from the fact that X is τ_1 and $\mathscr{G}$ is upper semicontinuous.) Let U and V be disjoint open sets in X separating

$\varphi^{-1}(g_1)$ and $\varphi^{-1}(g_2)$. Then U and V also contain open inverse sets U_I and V_I which separate $\varphi^{-1}(g_1)$ and $\varphi^{-1}(g_2)$, by upper semicontinuity. Now $\varphi(U_I)$ and $\varphi(V_I)$ are open sets in Q which give a Hausdorff separation of g_1 and g_2.

Now, if A and B are closed and disjoint subsets of Q, then $\varphi^{-1}(A)$ and $\varphi^{-1}(B)$ are closed and disjoint in X and can be separated by open sets U and V, which in turn will contain open inverse sets U_I and V_I separating $\varphi^{-1}(A)$ and $\varphi^{-1}(B)$. Thus $\varphi(U_I)$ and $\varphi(V_I)$ separate A and B in Q and are open, so that Q is normal.

If $\mathscr{G}$ is lower semicontinuous and $\{U_\alpha\}_{\alpha \in A}$ is a countable basis for the open sets in X, then $\{\varphi(U_\alpha)\}_{\alpha \in A}$ is a countable collection of open sets in Q, since φ is an open mapping. The map φ is open because $\mathscr{G}$ is lower semicontinuous. Then $\{\varphi(U_\alpha)\}$ is a basis for the open sets in Q, since any open set in Q has a corresponding open inverse set in X which can be expressed as a union of a subcollection of the U_α. Hence Q is perfectly separable.

If X is separable and metric and $\mathscr{G}$ is both upper and lower semicontinuous, then we see at once that Q is τ_1, since X is τ_1 and $\mathscr{G}$ is upper semicontinuous. Also Q is normal by (2), so Q is regular. Now Q is perfectly separable, since X is perfectly separable and $\mathscr{G}$ is lower semicontinuous by (3). But a regular, τ_1, perfectly separable space is metric, so Q is a metric space.

7. The quotient space Q of a decomposition $\mathscr{G}$ of a compact Hausdorff space X is itself Hausdorff if and only if $\mathscr{G}$ is upper semicontinuous.

If X is Hausdorff, then it is surely τ_1, so Q is τ_1, since $\mathscr{G}$ is upper semicontinuous. Hence the elements g of $\mathscr{G}$ are closed sets in X and compact. But disjoint compact sets can be separated by open sets in a Hausdorff space, so let g_1, g_2 lie in the disjoint open sets U and V. Then U and V each contain open inverse sets which separate g_1 and g_2 and whose images in Q separate the points g_1 and g_2.

Conversely, if X is compact and Hausdorff, and Q is Hausdorff, then φ, the natural decomposition function, is a closed mapping. If U is open in X, then $\varphi(X \sim U)$ is closed in Q and $\varphi(X \sim U)$ is the collection of g not contained in U. Hence if $G = \bigcup_{g \ \cup \ g}$, then $\varphi(G)$ is open in Q, so G must be open in X. Therefore $\mathscr{G}$ is upper semicontinuous.

EXAMPLE. The question may be asked, "Can compactness of X be weakened to local compactness in the statement of the preceding theorem?" The answer is negative, as demonstrated in this example. Let X be the subspace of E^2 defined by

$$X = \{(x,y) : y = 1, \tfrac{1}{2} < x \le 1, \text{ or}$$
$$y = \tfrac{1}{2}, 0 \le x \le \tfrac{3}{4}\}.$$

Let Y be the subspace of E^2 defined by

$$Y = \{(x,y) : y = 0, 0 \le x \le 1\}.$$

Define $f : X \to Y$ by $f(x,y) = (x,0)$ (i.e., projection on the first factor), and let $\mathscr{G}$ be the natural decomposition of f, and Q the quotient space of $\mathscr{G}$.

The quotient space is Hausdorff, since Q is just an interval. However, $\mathscr{G}$ is not upper semicontinuous, since the subset B of X is open in X, but the union of all elements of $\mathscr{G}$ contained in B is the closed interval $\{(x,y) : y = \tfrac{1}{2}, 0 \le x \le \tfrac{1}{2}\}$:

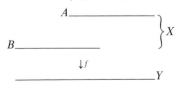

8. Let X be Hausdorff and $f: X \Rightarrow Y$ a quasicompact mapping with compact point-inverse frontiers. If the natural decomposition of f is upper semicontinuous, then Y is Hausdorff.

From the fact that X is Hausdorff and $\mathcal{G}$ is upper semicontinuous, we know that the elements g of $\mathcal{G}$ are closed sets, or equivalently, point inverses of f are closed. Also $\operatorname{int}(f^{-1}(y)) = f^{-1}(y) \sim \operatorname{Fr}(f^{-1}(y))$ is open, since its complement $\operatorname{Cl}(X \sim f^{-1}(y))$ is closed. Now let y_1 and y_2 be distinct points in Y. Then the sets $f^{-1}(y_1) \cap \operatorname{Cl}(X \sim f^{-1}(y_1))$ and $f^{-1}(y_2) \cap \operatorname{Cl}(X \sim f^{-1}(y_2))$ are disjoint and compact, so let F_1 and F_2 be disjoint open sets separating them. Then the open sets $\mathcal{F}_1$ and $\mathcal{F}_2$ also separate these point-inverse frontiers, where $\mathcal{F}_1 = F_1 \cap (X \sim f^{-1}(y_2))$ and $\mathcal{F}_2 = F_2 \cap (X \sim f^{-1}(y_1))$. Now the sets $U_1 = \mathcal{F}_1 \cup \operatorname{int}(f^{-1}(y_1))$ and $U_2 = \mathcal{F}_2 \cup \operatorname{int}(f^{-1}(y_2))$ are disjoint and open, and hence contain open inverse sets U_{I_1} and U_{I_2} such that $f(U_{I_1})$ and $f(U_{I_2})$ are disjoint open sets separating y_1 and y_2 in Y.

9. Given a monotone mapping $f: X \Rightarrow Y$ where X and Y are Hausdorff spaces and X is locally compact, then the natural decomposition of f is upper semicontinuous.

Let $\mathcal{G}$ be the decomposition of f, and denote the elements of $\mathcal{G}$ by g. It is quickly seen that each set g of $\mathcal{G}$ lies in an open set of compact closure. Now consider U, an open set in X. Let g be an arbitrary element of $\mathcal{G}$ contained in U. If we can get an open set about g which meets only those g's contained in U, and then cover $\bigcup_{g \subset U} g$ with such sets, then the union of these sets will equal $\bigcup_{g \subset U} g$, and $\mathcal{G}$ will be upper semicontinuous.

Accordingly, if $g_k \subset U$, an open set, let V be an open set with compact closure about g_k, and define $R = U \cap V$. Now let $C = \operatorname{Fr}(R)$. Then C and g_k are disjoint compact sets, so they lie in disjoint open sets, say Z_c and Z_g. Now $f(C)$ is closed in Y (since C is compact), so $f^{-1}(Y \sim f(C))$ is open in X. We claim that the set $P = R \cap Z_g \cap f^{-1}(Y \sim f(C))$ is an open set about g_k meeting only those g contained in U. For P can meet no set g which is not in R, since if g meets R and g meets $X \sim R$, then g must meet $\operatorname{Fr}(R)$, since g is connected. However, then g is contained in $f^{-1}(f(C))$, so g misses P, and P is the required open set.

10. If $f: X \Rightarrow Y$ is a quasicompact monotone mapping where X is locally compact and Hausdorff, then Y is Hausdorff if and only if the natural decomposition $\mathcal{G}$ of f is upper semicontinuous.

The necessity part was just proved. For the converse, we just note that point-inverses are compact and hence afford an open set separation in X. Then we use the upper-semicontinuity of $\mathcal{G}$ to get open inverse sets separating $f^{-1}(y_1)$ and $f^{-1}(y_2)$ in X and the quasi-compactness of f gives the result immediately.

11. If the decomposition $\mathcal{G}$ of a space X is upper semicontinuous and has compact element frontiers, and if X is separable and metric, so also is the decomposition space Q of $\mathcal{G}$.

We must show that Q is a normal, τ_1, perfectly separable space. From previous work we know that Q is normal and τ_1, since X is normal and $\mathcal{G}$ is upper semicontinuous. Now let $R = \{R_1, R_2, \ldots\}$ be a countable basis for the open sets in X. Then we assert that at most a countable number of the g in $\mathcal{G}$ have nonempty interiors. For, since the collection $\{\operatorname{int}(g)\}$ is disjoint, each element must contain at least one distinct set of R, so $\{\operatorname{int}(g)\}$ must be countable. Now let $\mathcal{P} = \{\text{finite unions of sets from } R \text{ and } I\}$, where I is the collection of interiors of the sets g. We claim that $\mathcal{L} = \{\bigcup_{g \subset P} g : P \in \mathcal{P}\}$ forms a basis for the open sets in Q.

Let $p \in Q$, $p \in U$ an open set in Q. Then $\varphi^{-1}(p) = g_p$ lies in the open set $\varphi^{-1}(U)$ in X, so cover $\mathrm{Fr}(g_p)$ with a finite number of open subsets of $\varphi^{-1}(U)$ from R. Then, if

$$g_p \subset \left[\bigcup_{i=1}^{N} R_i \cup \mathrm{int}(g_p) \right] = V,$$

since V is open and $\mathscr{G}$ is upper semicontinuous, it follows that the inverse set $G = \bigcup_{g \subset V} g$ is open, and $\varphi(G)$ is open in Q, contains p, and is contained in U. But $G \in \mathscr{L}$, so $\mathscr{L}$ is a countable basis for Q.

Hence Q is a metrizable topological space. To show that Q is separable, let $\{x_i\}_{i=1}^{\infty}$ be a dense set in X. Then the set $\{\varphi(x_i)\}_{i=1}^{\infty}$ is dense in Q, for if U is any open set in Q which misses the set $\{\varphi(x_i)\}_{i=1}^{\infty}$, then $\varphi^{-1}(U)$ is open in X and misses the set $\{x_i\}_{i=1}^{\infty}$, which is impossible.

SECTION III

Component Decomposition

Definition. For any function $f: X \Rightarrow Y$ the decomposition of X into components of point inverses is called the *component decomposition of f*.

EXAMPLES.

(1) The component decomposition and the natural decomposition of a function f coincide if f is a monotone mapping.
(2) The component decomposition of the mapping of $[0,2\pi]$ onto the unit circle by $f(x) = e^{ix}$ consists of the decomposition of $(0,2\pi)$ into individual points, and the points 0 and 2π are components of the point inverse $f^{-1}f(0)$. The component and natural decompositions differ only at $f^{-1}f(0)$.

Notational Convention. We term the component decomposition $\mathcal{M}$, its quotient space M, and its elements m (either sets in X or points in M).

Definition. A function $l: X \Rightarrow Y$ is said to be *light* provided that for each $y \in Y$, $l^{-1}(y)$ is totally disconnected, that is, the components of $l^{-1}(y)$ are single points.

Definition. If $f: X \to Y$ is a mapping, then a *factorization* of f is any representation $f = g \circ \varphi$ where $\varphi: X \Rightarrow M$, $g: M \to Y$ are mappings; X, M, and Y are topological spaces; and $f(x) = (g \circ \varphi)(x)$ for each x in X.

Definition. Two factorizations $f = g_1 \circ \varphi_1$ and $f = g_2 \circ \varphi_2$ of a mapping $f: X \to Y$ are *topologically equivalent* provided that there exists a homeomorphism $h: M_1 \Rightarrow M_2$ such that $\varphi_2 = h \circ \varphi_1$ and $g_1 = g_2 \circ h$.

Definition. A factorization of a mapping is *topologically unique* provided that it is topologically equivalent to any other factorization of the mapping.

EXAMPLE (A factorization which is not topologically unique). Let $Y = I$, and X be a discrete subspace of I. Define f to be the identity projection of X into Y, i_X the identity map of X, and i_Y the identity map of Y. Then $f = f \circ i_X$ and $f = i_y \circ f$. The first factors are monotone, the second factors are light, but the factorizations are clearly not topologically equivalent. This situation cannot arise if we require that f be quasicompact, as we will see later.

EXAMPLE. Let X be the unit sphere minus the xy plane, and f the projection defined by sending the point (x,y,z) in X to the point $(x,0,0)$. Then the component decomposition is homeomorphic to two disjoint open intervals (i.e., $\mathcal{M} = \{(x,y) : y = 0, 0 < x < 1\} \cup \{(x,y) : y = 1, 0 < x < 1\}$).

EXERCISES III.

1. Let $f : X \Rightarrow Y$ be a mapping with compact components of point inverses, where X, Y are Hausdorff and X is locally compact. Then the component decomposition is upper semicontinuous. Also:

   (1) The quotient space M is Hausdorff.
   (2) The natural mapping $\varphi : X \Rightarrow M$ is closed and monotone.
   (3) M is locally compact, and M is normal if X is normal.

2. Given Hausdorff spaces X and Y with X locally compact, any mapping $f : X \Rightarrow Y$ with compact components of point inverses admits a topologically unique factorization of $f = l \circ \varphi$, where:

   (1) $\varphi : X \Rightarrow M$ is closed and monotone;
   (2) $l : M \Rightarrow Y$ is light, and M is locally compact and Hausdorff;
   (3) if X is separable and metric, so also is M.

SOLUTIONS.

1. Let $f: X \Rightarrow Y$ be a mapping with compact components of point inverses, where X, Y are Hausdorff and X is locally compact. Then the component decomposition is upper semicontinuous. Also:

(1) The quotient space M is Hausdorff.
(2) The natural mapping $\varphi : X \to M$ is closed and monotone.
(3) M is locally compact, and M is normal if X is normal.

In proving this exercise, we make use of the following lemma.

Lemma. *Let K be a closed subset of X, a locally compact Hausdorff space. If the components of K are compact and U is an open set about m, a component of K, then there exists an open set V with compact closure such that $\mathrm{Fr}(V) \cap K = \varnothing$ and $m \subset V \subset \mathrm{Cl}(V) \subset U$.*

Since $m \subset U$ is compact and X is locally compact, we know we have an open set W containing m such that $\mathrm{Cl}(W) \subset U$ and $\mathrm{Cl}(W)$ is compact. Suppose $\mathrm{Fr}(W) \cap K \neq \varnothing$. Then $\mathrm{Cl}(W) \cap K$ is a Hausdorff subspace, and no component of $\mathrm{Cl}(W) \cap K$ meets both m and $\mathrm{Fr}(W) \cap K$, so we use the result of Exercise 2 in Part I, §XI to get a separation of $\mathrm{Cl}(W) \cap K = A \cup B$ with $m \subset A$, $\mathrm{Fr}(W) \cap K \subset B$.

Now the pair A and $B \cup \mathrm{Fr}(W)$ are disjoint compact sets in $\mathrm{Cl}(W)$, and hence we have disjoint sets V and Y, open in $\mathrm{Cl}(W)$, with $A \subset V$ and $B \cup \mathrm{Fr}(W) \subset Y$. Clearly V is open in X [as $V \cap \mathrm{Fr}(W) = \varnothing$], and $\mathrm{Fr}(V) \cap K = \mathrm{Fr}(V) \cap [\mathrm{Cl}(W) \cap K] = \mathrm{Fr}(V) \cap [A \cup B] = \varnothing$, since V contains A and Y contains B. Therefore V is the required set.

Since Y is T_1, point inverses are closed in X. Let U be an open set in X and $m \subset U$. If we can get an open set about m which meets only other m lying wholly in U, for each $m \subset U$, then we shall have shown that $\bigcup_{m \subset U} m$ is open and $\mathscr{M}$ is upper semicontinuous. Suppose $m \subset U$ is a component of $f^{-1}(y)$, a closed set. Then, applying the lemma, we get an open set V about m with $\mathrm{Cl}(V) \subset U$ and $\mathrm{Fr}(V) \cap f^{-1}(y) = \varnothing$. Now let $C = \mathrm{Fr}(V)$, a compact set. Hence $f(C)$ is closed in Y, since Y is Hausdorff, so $f^{-1}(Y \sim f(C))$ is open in X. Also, since $f^{-1}(y) \cap C$ is void, then m, a component of $f^{-1}(y)$, lies in $f^{-1}(Y \sim f(C))$. Because each m is connected, any m which meets V and $X \sim V$ must meet $\mathrm{Fr}(V) = C$, so the set $V \cap f^{-1}(Y \sim f(C))$ meets only those m lying inside V. This, then, is the required open set about m, so we may conclude $\bigcup_{m \subset U} m$ is open and $\mathscr{M}$ is upper semicontinuous.

To show that M is Hausdorff, take m_1, m_2 distinct in M. Then $\varphi^{-1}(m_1)$ and $\varphi^{-1}(m_2)$ are compact, disjoint sets in X, a Hausdorff space, so let U, V be open sets separating them. Then $U_I = \bigcup_{m \subset U} m$ and $V_I = \bigcup_{m \subset V} m$ are open inverse sets relative to φ, and $\varphi(U_I)$, $\varphi(V_I)$ separate m_1, m_2 in M.

That φ is monotone follows by definition. That φ is closed follows because φ is quasicompact and $\mathscr{M}$ is upper semicontinuous [see Exercise 6(b), §II].

To verify the local compactness of M, let $p \in M$. Then $\varphi^{-1}(p)$ lies in an open set O with compact closure. Then the set $W = \bigcup_{m \subset O} m$ is an open set about $\varphi^{-1}(p)$ with compact closure, so $\varphi(W)$ is open in M, contains p, and has compact closure. If X is normal, M is normal; for let B_1 and B_2 be disjoint closed sets in M. Then $\varphi^{-1}(B_1)$ and $\varphi^{-1}(B_2)$ are closed in X and disjoint, so they can be separated by open disjoint sets $\mathscr{B}_1$ and $\mathscr{B}_2$, each of which contains an open inverse set whose image under φ yields an open set separation of B_1 and B_2 in M.

2. Given Hausdorff spaces X and Y with X locally compact, any mapping $f: X \Rightarrow Y$ with compact components of point inverses admits a topologically unique factorization $f = l \circ \varphi$, where:

(1) $\varphi: X \Rightarrow M$ is closed and monotone;
(2) $l: M \Rightarrow Y$ is light, and M is locally compact and Hausdorff;
(3) if X is separable and metric, so also is M.

From the previous theorem we know that M is locally compact and Hausdorff, and that $\varphi: X \Rightarrow M$ is closed and monotone. We also showed that the decomposition $\mathcal{M}$ of X is upper semicontinuous, and $\mathcal{M}$ has compact element frontiers, since $\mathcal{M}$ has compact elements. Then we apply Exercise 11, §II to prove that M is separable and metric if X is. The following lemma will be useful here to prove that l is light.

Lemma. *Let $f: X \Rightarrow Y$ be a closed, monotone mapping where X and Y are topological spaces. Then if S is a closed connected set in Y, $f^{-1}(S)$ is connected in X.*

PROOF. We assume that $f^{-1}(S) = M_1 \cup M_2$ is a separation in X. Both M_1 and M_2 are closed in X so $f(M_1)$ and $f(M_2)$ are closed, since f is a closed mapping. Now we assert that $f(M_1) \cap f(M_2) = \varnothing$. For, if $f^{-1}(y)$ meets M_1 and M_2 in X, suppose $f^{-1}(y) \cap M_1 = I_1$ and $f^{-1}(y) \cap M_2 = I_2$. Then I_1 and I_2 are each closed so $f^{-1}(y) = I_1 \cup I_2$—a contradiction, since f is monotone. Thus we have $S = f(M_1) \cup f(M_2)$, a separation, which contradicts the connectedness we assumed for S.   □

Suppose that $l^{-1}(y)$ contains a nondegenerate component S in M. Note that S is closed in M, since it is closed in $l^{-1}(y)$, and recall that φ is closed and monotone. Then, applying the lemma, we see that $\varphi^{-1}(S)$ is connected in X, and since $\varphi^{-1}(S) \subset f^{-1}(y)$, we see that S must be a single point in M. Therefore $l: M \Rightarrow Y$ is light.

Now it remains to verify that $f = l \circ \varphi$ is a topologically unique factorization. Suppose that $f = l_1 \circ \varphi_1$, where $\varphi_1: X \Rightarrow M_1$ is closed and monotone, l_1 is light, and M_1 is locally compact and Hausdorff. Our objective is to show that there is a homeomorphism h from M_1 onto M such that $h \circ \varphi_1 = \varphi$ and $l_1 = l \circ h$. (See Figure III.1.)

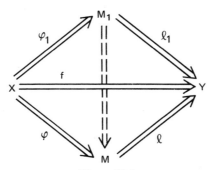

Figure III.1

We define $h = \varphi \circ \varphi_1^{-1}$; $h: M_1 \Rightarrow M$. First we show that h is well defined. Let m be a point of M_1. Then $\varphi_1^{-1}(m) \subset f^{-1}(y_0)$ for some y_0 in Y. Therefore, since $h(m) = \varphi \circ \varphi_1^{-1}(m)$ is a connected subset of $l^{-1}(y_0)$ in M_2, and l is light, then $h(m)$ is a single point. We use a similar argument on the lightness of l_1 to show h^{-1} is well defined. Clearly h is onto, since $\varphi: X \Rightarrow M$ is onto.

Now let S be any set in M_1. Then we claim that $\varphi_1^{-1}(S)$ is an inverse set in X relative to φ. If not, there is an m_0 in M such that $\varphi^{-1}(m_0)$ meets $\varphi_1^{-1}(S)$ in X but is not contained in it. However, $m_0 \in l^{-1}(y_0)$ for some y_0 in Y, and $\varphi^{-1}(m_0)$ is connected in X, so $\varphi_1(\varphi^{-1}(m_0))$ is connected in M_1 and is a subset of $l_1^{-1}(y_0)$, and hence a single point. However, $\varphi_1(\varphi^{-1}(m_0))$ meets S, so it must be a point of S, and thus $\varphi_1^{-1}(S)$ contains $\varphi^{-1}(m_0)$. Therefore $\varphi_1^{-1}(S)$ is a φ-inverse set in X. Now we may quickly see that $\varphi^{-1}(S)$ is a φ_1-inverse set in X for any set S in M, by a similar argument.

Hence let S be open in M. Then $h^{-1}(S) = \varphi_1 \circ \varphi^{-1}(S)$ is open, since φ is continuous and φ_1 is quasicompact. Reversing these roles, we see that h is an open mapping, so we have the desired homeomorphism.

Homotopy

Definition. Given two mappings $f, g : X \to Y$, f is said to be *homotopic* to g ($f \simeq g$) provided that there exists a mapping $h : X \times I \to Y$ with $h(x,0) = f(x)$ and $h(x,1) = g(x)$ for each x in X.

Remarks. We may regard h as a mapping from the cylinder $X \times I$ into Y. We also may regard the image $h(X,t)$ in Y as being deformed continuously from $f(X)$ at time $t = 0$ to $g(X)$ at time $t = 1$. If $f : X \to Y$ is homotopic to a constant mapping, then $f(X)$ may be shrunk continuously to a single point in Y. If X is a compact metric space, then h is a uniformly continuous mapping.

EXERCISES IV.

1. Homotopy is an equivalence relation.

2. Any mapping $f : X \to E^N$ is homotopic to a constant,

3. Any two mappings of X into E^N are homotopic to each other.

4. The property of a space Y whereby any two mappings of a given space X into Y are homotopic to each other is a topological invariant.

5. For any p in S^n (the n-dimensional sphere), the space $S^n \sim p$ (the punctured n-sphere) has the property that any mapping of a space X into $S^n \sim p$ is homotopic to a constant mapping.

6. If $h : S^1 \to S^1$ is a homeomorphism, then h is not homotopic to a constant mapping.

7. A mapping $f : S^1 \to Y$ is homotopic to a constant mapping if and only if there exists a continuous extension F of f to the disk bounded by S^1 in the plane.

8. Given a mapping $f : C \to S^n$, where C is a closed subset of a metric space X, there exists an open set U in X containing C and a continuous extension F of f to U, $F : U \to S^n$.

SOLUTIONS.

1. Homotopy is an equivalence relation.

Let f, g, k be mappings of X into Y. Evidently $f \simeq f$, since $h(x,t) = f(x)$ is continuous for all t in $[0,1]$. Now if $f \simeq g$, we have $h : X \times I \to Y$, so $h'(x,t) = h(x, 1 - t)$ is a mapping from $X \times I$ to Y with $h'(x,0) = g(x)$, $h'(x,1) = f(x)$.

Finally suppose $f \simeq g$ and $g \simeq k$, so we have the mappings $h_1, h_2 : X \times I \to Y$ with $h_1(x,0) = f(x)$, $h_1(x,1) = g(x)$, $h_2(x,0) = g(x)$, and $h_2(x,1) = k(x)$. Then the function $h'(x,t) = h_1(x,2t)$ is a mapping from $X \times [0,\frac{1}{2}]$ into Y, and $h''(x,t) = h_2(x,2t - 1)$ is a mapping from $X \times [\frac{1}{2},1]$ into Y; furthermore, h' and h'' agree on $X \times \{\frac{1}{2}\}$. Hence the function h defined by

$$h(x,t) = \begin{cases} h'(x,t) & \text{if } 0 \le t \le \frac{1}{2}, \\ h''(x,t) & \text{if } \frac{1}{2} \le t \le 1 \end{cases}$$

is a mapping from $X \times I$ into Y with $h(x,0) = f(x)$ and $h(x,1) = k(x)$, so that $f \simeq k$.

2. Any mapping $f : X \to E^N$ is homotopic to a constant.

Suppose $f : X \to E^N$ is a mapping. Then, since E^N is a convex space (that is, every point on a straight line segment between two points of the space is itself in the space), the function $h(x,t) = (1 - t)f(x) + kt$ is defined and continuous for any point k in E^N. If $k(x) = k$ is the constant mapping of X into E^N corresponding to the point k, then since $h(x,0) = f(x)$ and $h(x,1) = k = k(x)$, h is a homotopy from f to k.

3. Any two mappings of X into E^N are homotopic to each other.

Just note that if f, g are mappings of X into E^N and $f \simeq k_1$, $g \simeq k_2$, then $k_1 \simeq k_2$ by $h(x,t) = (1 - t)k_1 + tk_2$. Using the fact that homotopy is an equivalence relation, we have $f \simeq g$.

4. The property of a space Y whereby any two mappings of a given space X into Y are homotopic to each other is a topological invariant of the space Y.

Suppose that X is a given space and Y has the property mentioned. Then let $s : Y \to Z$ be a homeomorphism. We claim that Z has this property also.

If f, g are mappings of X into Z, then $f' = s^{-1} \circ f$ and $g' = s^{-1} \circ g$ are mappings of X into Y, and hence $f' \simeq g'$:

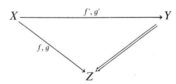

If $h(x,t)$ is the homotopy from f' to g' $[h(x,0) = f'(x)$ and $h(x,1) = g'(x)]$, then $h' = s \circ h$ is a mapping from $X \times I$ into Z with $h'(x,0) = s \circ f'(x) = f(x)$ and $h'(x,1) = s \circ g'(x) = g(x)$, so $f \simeq g$.

5. For any point p in S^n (the n-dimensional sphere), the space $S^n \sim p$ (the punctured n-sphere) has the property that any mapping of a space X into $S^n \sim p$ is homotopic to a constant mapping.

We know that the space E^n has the property that any mapping of a space X into E^n is homotopic to a constant. Thus, according to the above exercise, we need only show that E^n is homeomorphic to $S^n \sim p$ in order to be finished. We will establish such a homeomorphism for the case $n = 2$.

Consider the unit sphere with a rectangular coordinate system whose origin coincides with the sphere's center (Figure IV.1). Take p to be the point $(0,0,1)$. Our homeomorphism will associate the point (x,y,z) in $S^2 \sim p$ with the point $(x',y',0)$ in the plane $z = 0$, where (x,y,z) is the point of intersection of a straight line between $(0,0,1)$ and $(x',y',0)$.

It is easily verified that $F : S^2 \sim p \Rightarrow E^2$ defined by

$$F(x,y,z) = \left(\frac{x}{1-z}, \frac{y}{1-z}, 0 \right)$$

is continuous. Now $F^{-1} : E^2 \Rightarrow S^2 \sim p$ is given by

$$F^{-1}(u,v,0) = \left(\frac{2u}{u^2 + v^2 + 1}, \frac{2v}{u^2 + v^2 + 1}, \frac{u^2 + v^2 - 1}{u^2 + v^2 + 1} \right).$$

That $F^{-1} \circ F$ and $F \circ F^{-1}$ are the identity maps can be easily checked and F^{-1} is continuous, so $S^2 \sim p \cong E^2$ and we are finished.

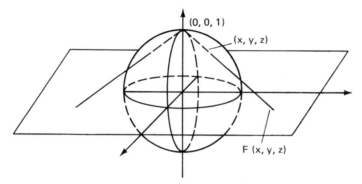

(0, 0, 1)

(x, y, z)

F (x, y, z)

Figure IV.1

Lemma. *If f_1, f_2 are mappings from a space X into S^1 such that $f_1(x) = e^{iu(x)}$ and $p(f_1(x),f_2(x)) < 2$ for all x, then $f_2(x) = e^{i[u(x) + v(x)]}$, where $v(x) = [f_2(x),f_1(x)]$ is a signed angle.*

PROOF. Let x be fixed. Then if $f_1(x) = e^{i\theta_1}$, $f_2(x) = e^{i\theta_2}$, we have $v(x) = \theta_2 - \theta_1$ and $e^{i(\theta_2 - \theta_1)} = e^{i\theta_2}/e^{i\theta_1} = f_2(x)/f_1(x)$. Hence

$$e^{i[u(x) + v(x)]} = e^{iu(x)}e^{iv(x)} = f_1(x)e^{iv(x)} = f_1(x)[f_2(x)/f_1(x)] = f_2(x). \qquad \square$$

Definition. A mapping f of a topological space X into the circle S^1 is said to be *exponentially representable* provided that f admits a factorization $f = p \circ u$, where $u : X \to R^1$ is a mapping and $p : R^1 \to S^1$ is the mapping $p(x) = e^{ix}$.

Lemma. *If a mapping $f : X \to S^1$, X a compact metric space, is homotopic to a constant, then f is exponentially representable.*

PROOF. Suppose that $c \simeq f$. Then the homotopy $h : X \times I \to S^1$ is uniformly continuous, since $X \times I$ is compact. Choose $\varepsilon > 0$ such that $|h(x_1,t_1) - h(x_2,t_2)| < 2$ whenever $p((x_1,t_1),(x_2,t_2)) < \varepsilon$, and partition I with the points $0 = t_0, t_1, \ldots, t_{N-1}, t_N = 1$ such that $p(t_i,t_{i+1}) = |t_i - t_{i+1}| < \varepsilon$.

Now, since $h(x,0) = c$ for all $x \in X$, for any x we have $|h(x,t) - c| < 2$ for $t \in]0,t_1]$. In particular, $|h(x,t_1) - c| < 2$ for all $x \in X$. Since $c = e^{i\theta_0}$ for some θ_0, the constant function c is exponentially representable, so that by the previous lemma, $h : X \times \{t_1\} \to S^1$ is exponentially representable.

Applying the lemma again, we have that $h : X \times \{t_2\} \to S^1$ is exponentially representable. After finitely many repetitions of this argument, we conclude that $h : X \times \{t_N\} \to S^1$ is exponentially representable. But $h(x,t_N) = f(x)$, so the lemma is proved.   □

6. If $h : S^1 \to S^1$ is a homeomorphism, then h is not homotopic to a constant mapping.

Since S^1 is a compact metric space, then h has an exponential representation $h(x) = e^{iu(x)}$ if h is homotopic to a constant. Then $u(S^1)$ is a continuum and hence a compact interval in R^1. Since $h = p \circ u$, u must be 1-1. But a 1-1 map from a compact Hausdorff space to a Hausdorff space is a homeomorphism. This implies that S^1 is homeomorphic to an interval, which is a contradiction, since an interval has cut points and S^1 does not.

7. A mapping $f : S^1 \to Y$ is homotopic to a constant mapping if and only if there exists a continuous extension F of f to the disk bounded by S^1 in the plane.

Define $\psi : S^1 \times I \to D^2$ (disk bounded by S^1) by $\psi(e^{i\theta},t) = (1 - t)e^{i\theta}$. Then ψ is continuous; ψ is also quasicompact, since if U is any open inverse set in $S^1 \times I$, then U^c is compact, so that $\psi(U^c)$ is compact and hence closed. But $\psi(U) = [\psi(U^c)]^c$, so $\psi(U)$ is open.

Now, suppose that $f : S^1 \to Y$ has a continuous extension $F : D^2 \to Y$. Consider $h = F \circ \psi$. Clearly h is continuous; also, since F is an extension, $h(x,0) = F(\psi(x,0)) = F(x) = f(x)$ and $h(x,1) = F(\psi(x,1)) = F(0)$. Therefore h is a homotopy between f and k, where k is the constant map defined by $k(x) = F(0)$.

Conversely, if h is a homotopy from f to a constant map, let $F : D^2 \to Y$ be given by $F = h \circ \psi^{-1}$. Since $\psi^{-1}(x)$ is a single point for $x \neq 0$ and h is constant on $\psi^{-1}(0)$, F is well defined. If U is an open set in Y, then $F^{-1}(U)$ is open if and only if $\psi^{-1} \circ F^{-1}(U)$ is open, since ψ is quasicompact. But $\psi^{-1} \circ F^{-1} = (F \circ \psi)^{-1} = h^{-1}$ by construction, and h is continuous. Therefore F is continuous, and since $F(e^{i\theta}) = h \circ \psi^{-1}(e^{i\theta}) = h(e^{i\theta},0) = f(e^{i\theta})$, F is the required extension.

To complete this section, we need the following theorem on extensions of mappings into R^1:

Tietze Extension Theorem. *Given any real-valued continuous function $f : C \to R^1$ where C is a closed subset of metric space X such that $|f(x)| \leq M$ on C, there exists a continuous extension $F : X \to R^1$ such that $|F(x)| \leq M$ on X.*

PROOF. We will construct a uniformly convergent sequence of continuous functions and take $F(x)$ to be the limit. First we construct $F_1(x)$ with $|F_1(x)| \leq M/3$ on X and $|f(x) - F_1(x)| \leq 2M/3$ on C. Define $C^- = \{x \in C : f(x) \leq -M/3\}$ and similarly $C^+ = \{x \in C : f(x) \geq M/3\}$. Now let F_1 be defined by

$$F_1(x) = \frac{M}{3} \left\{ \frac{p(x,C^-) - p(x,C^+)}{p(x,C^-) + p(x,C^+)} \right\},$$

where $\rho(x,\varnothing) = 1$ by definition. Then $F_1(x) = -M/3$ on C^- and $F_1(x) = M/3$ on C^+, and $|F_1(x)| \leq M/3$ on X. By examining the three cases $x \in C^-$, $x \in C^+$, and $x \in X \sim (C^- \cup C^+)$, one can easily verify that $|f(x) - F_1(x)| \leq 2M/3$.

Next replace $f(x)$ by $f(x) - F_1(x)$ and M by $2M/3$ to obtain a map F_2 such that $|F_2(x)| \leq 2M/3^2$ on X and $|f(x) - F_1(x) - F_2(x)| \leq (\frac{2}{3})^2 M$. Similarly get F_3 with $|F_3(x)| \leq 2^2 M/3^3$ and $|f(x) - F_1(x) - F_2(x) - F_3(x)| \leq (\frac{2}{3})^3 M$.

The general term F_n has the property that $|F_n(x)| \leq 2^{n-1} M/3^3$ on X and

$$\left| f(x) - \sum_{k=1}^{n} F_k(x) \right| \leq \left(\frac{2}{3}\right)^n M.$$

Define $F(x) = \sum_{k=1}^{\infty} F_k(x)$; then $F(x)$ is a continuous extension of $f(x)$ with

$$|F(x)| \leq \sum_{n=1}^{\infty} \frac{2^{n-1} M}{3^n}$$

$$= \frac{M}{3} \left(\frac{1}{1 - \frac{2}{3}} \right) = M$$

on X.   □

8. Given a mapping $f: C \to S^n$, where C is a closed subset of the metric space X, there exists an open set U in X containing C and a continuous extension F of f to U, $F: U \to S^n$.

Consider the range space S^n as a subspace of E^{n+1}, and note that the projecting functions yield continuous maps $p_i \circ f : C \to E_i$. By the Tietze theorem, each such mapping may be extended to a mapping $F_i: X \to E_i$. Then the function G defined by $G(x) = (F_1(x), F_2(x), \ldots, F_{n+1}(x))$ is an extension of f. Consider $U = X \sim G^{-1}(0)$; U is open and contains C. Define $F(x) = G(x)/|G(x)|$ for $x \in U$. Then F is clearly a continuous map of U into S^n and extends f.

A result which we incorporate into this section and will need for later application is the following:

Homotopy Extension Theorem. *Given mappings $f, g : C \to S^n$ where C is a closed subset of a metric space X with $f \simeq g$ on C and an extension $F : X \to S^n$ of f. Then there exists an extension $G : X \to S^n$ of g such that $F \simeq G$ on X.*

PROOF. Suppose h is a homotopy with $h(x,0) = f(x)$ and $h(x,1) = g(x)$ on $C \times I$. Define a function F by

$$F(x,t) = \begin{cases} F(x) & \text{if } t = 0 \\ h(x,t) & \text{if } (x,t) \in C \times I \end{cases}$$

Then F maps $X \times \{0\} \cup C \times I$ into S^n, since it is the combination of two mappings which agree on the common closed $C \times \{0\}$. By Exercises 8 of this section, F may be extended to an open set U about $X \times \{0\} \cup C \times I$.

We wish to show that there is an open set V in X such that $C \subset V$ and $V \times I \subset U$. Given any point $(x,t) \in C \times I$, we can find a neighborhood of x in X, $U_t(x)$, and $\varepsilon(x,t) > 0$ such that $U_t(x) \times (t - \varepsilon(x,t), t + \varepsilon(x,t)) \subset U$. For a fixed $x_0 \in C$, the open intervals $(t - \varepsilon(x_0,t), t + \varepsilon(x_0,t))$, $t \in I$, cover I, and by I's compactness we may extract a finite subcover, say $(t_1 - \varepsilon(x_0,t_1), t_1 + \varepsilon(x_0,t_1)), \ldots, (t_n - \varepsilon(x_0,t_n), t_n + \varepsilon(x_0,t_n))$. We then define $N(x_0) = \bigcap_{i=1}^{n} U_{t_i}(x_0)$ and note that this is a neighborhood of x_0 with the property that $N(x_0) \times I \subset U$. Now let $V = \bigcup_{x \in C} N(x)$, the union of the neighborhoods obtained by the above procedure; it is then easily verified that V is an open set in X which contains C and which has the property that $V \times I \subset U$.

Next we use the Urysohn mapping $U: X \to I$ such that

$$u(x) = \begin{cases} 0 & \text{if } x \in X \sim V, \\ 1 & \text{if } x \in C. \end{cases}$$

Now if we define $H(x,t) = F(x, t \cdot u(x))$, then H is defined and continuous on $X \times I$ and $H|C \times I = h(x,t)$, while $H(x,0) = F(x)$. Taking $G(x) = H(x,1)$, we see that G extends g to $X \times \{1\}$, and thus $H(x,t)$ is the required homotopy relating F and G. $\qquad\square$

Unicoherence

Definition. A connected space is *unicoherent* provided that, no matter how it is represented as the union of two closed, connected sets, the intersection of these sets is connected.

Definition. A *dendrite* is a locally connected generalized metric continuum containing no simple closed curve.

EXAMPLES.

(1) The circle is not unicoherent.
(2) A dendrite is unicoherent.

Definition. A connected metric space X has the *universal exponential represen-tation property* provided that every uniformly continuous function $f : X \to S^1$ can be represented in the form $f(x) = e^{iu(x)}$, where $u : X \to E^1$ is continuous.

Note. If $f(x) = e^{iu(x)}$ is exponentially representable and X is connected, then $u(x)$ is determined uniquely up to an additive constant. For if $f(x) = e^{iu(x)} = e^{iv(x)}$ then $e^{i(u(x)-v(x))} = 1$, so $u(x) - v(x) = 0, \pm 2\pi i, \pm 4\pi i, \ldots$, and since X is connected, $u - v$ must map X into one point of this discrete set, so $u(x) = v(x) + 2k\pi i$.

Definition. A *weakly monotone* mapping is a mapping whose point inverses are connected.

Lemma. *If $f : X \to Y$ is a weakly monotone, quasicompact mapping and S is closed and connected in Y, then $f^{-1}(S)$ is closed and connected in X.*

111

PROOF. It is immediate that $f^{-1}(S)$ is closed, since f is continuous. Now if $f^{-1}(S) = M_1 \cup M_2$ is a separation, then M_1 and M_2 are closed in X. We assert that M_1 and M_2 must be inverse sets, for if not, $f^{-1}(y)$ meets M_1 and M_2 for some y in S, and this is not possible, since $f^{-1}(y)$ is connected. Hence $f(M_1)$ and $f(M_2)$ are closed and disjoint. Thus S has a separation $S = f(M_1) \cup f(M_2)$, which is a contradiction. $\square$

Definition. *Given a metric space X with a distance function ρ, the metric determined by a distance function ρ' is said to be equivalent to ρ provided that the metric space X' determined by applying ρ' to X is homeomorphic to X.*

EXAMPLE. Let X be the unit disk, punctured at the center. Let ψ be the transformation which maps X onto the space $S^1 \times I_1$ by the action $\psi(z,\theta) = (1,\theta,z)$, where $I_1 = [0,1)$. Then the metric ρ_2 determined by $\rho_2 = \rho_1 \circ \psi$, where ρ_1 is the standard metric on $\mathbf{R}^3$, is an equivalent metric to the standard metric on X. (See Figure V.1.)

An equivalent metric on a metric space X is alternatively termed an admissible metric.

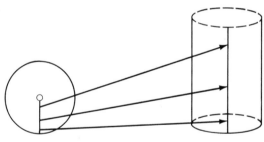

Figure V.1

EXERCISES V.

1. If $f: X \to Y$ is a weakly monotone, quasicompact function and X is connected and unicoherent, so also is Y.

2. Any multicoherent (nonunicoherent) connected metric space X admits a uniformly continuous (in an equivalent metric) mapping $f: X \to S^1$ which is not exponentially representable.

3. A locally connected metric continuum X is unicoherent if and only if every mapping $f: X \to S^1$ is exponentially representable.

4. If X_1, X_2 are each closed in their union $X = X_1 \cup X_2$, and if $X_1 \cap X_2$ is nonempty and connected, then if $f: X \to S^1$ is exponentially representable on X_1 and X_2, f is exponentially representable on X.

5. Let $X = \bigcup_{n=1}^{\infty} X_n = \bigcup_{n=1}^{\infty} \text{int}(X_n)$ be a metric space, where $X_1 \subset X_2 \subset X_3 \subset \cdots$ and each X_n is connected. Then if $f: X \to S^1$ is exponentially representable on each X_n, it is also on X.

6. A metric space X has the universal exponential representation property if for each $\varepsilon > 0$ and each admissible metric, $X = \bigcup_{n=1}^{\infty} D_n$, where each D_n is a closed connected set of diameter $< \varepsilon$ such that $D_{n+1} \cap (\bigcup_{i=1}^{n} D_i)$ is connected, and each x in X is interior to some finite union of these sets D_n. Then X is unicoherent and locally connected.

7. I^n, E^n $(n > 0)$, and S^n $(n > 1)$ have the universal exponential representation property and thus are unicoherent.

8. Every dendrite $\mathcal{D}$ has the universal exponential representation property.

9. If X is a connected, locally connected, unicoherent space, the boundary of each component of the complement of a closed and connected set K in X is itself connected.

10. If the boundary of each component of the complement of a closed, connected set in a locally connected and connected regular space X is itself connected, then X is unicoherent.

SOLUTIONS.

1. If $f : X \Rightarrow Y$ is a weakly monotone, quasicompact mapping and X is connected and unicoherent, so also is Y.

Since f is continuous, Y is connected. Now let A, B be a pair of closed connected subsets of Y whose union $A \cup B = Y$. By the lemma proved earlier in this section, $A_1 = f^{-1}(A)$ and $B_1 = f^{-1}(B)$ are closed and connected in X, so $A_1 \cap B_1$ is connected, and hence $f(A_1 \cap B_1) = f(f^{-1}(A) \cap f^{-1}(B)) = A \cap B$ is connected in Y.

2. Any multicoherent (nonunicoherent) connected metric space admits a uniformly continuous (in an equivalent metric) mapping $f : X \to S^1$ which is not exponentially representable.

Let A and B be two closed and connected sets whose union is X and whose intersection has the separation $A \cap B = M \cup N$. Then since M and N are each closed and disjoint, we have $\rho(M,N) > 0$, either in our original metric or in some equivalent metric. Therefore the function

$$f(x) = \frac{\rho(x,M)}{\rho(x,M) + \rho(x,N)}$$

is uniformly continuous on X. Now we define

$$h(x) = \begin{cases} e^{i\pi f(x)} & \text{if } x \in A, \\ e^{-i\pi f(x)} & \text{if } x \in B. \end{cases}$$

Then $h : X \to S^1$ is a uniformly continuous mapping, and we assume that $h(x)$ has an exponential representation. If so, then $h(x) = e^{iu(x)}$, where $u : X \to R^1$ is a mapping. Hence on the connected set A, $e^{i\pi f(x)} = e^{iu(x)}$ means that $\pi f(x) - u(x) = 2m\pi$ for an integer m. Also on B, $\pi f(x) + u(x) = 2n\pi$, since B is connected. Hence $f(x) = n + m$ on $A \cap B$.

However, on the set M, $f(x) = 0$, so $n = -m$. On the set N, $f(x) = 1$, so we get that $1 = m + n$ or $1 = 0$, a contradiction. Hence $h : X \to S^1$ is not exponentially representable.

Remark. The necessity of switching to an equivalent metric may not be appreciated. It is natural to ask: if X is a multicoherent connected metric space, can we not find two closed connected sets A and B such that $A \cup B = X$ and $A \cap B = M \cup N$ with $\rho(M,N) > 0$?

The answer is no, for the unit disk with the center removed is such a space, and there there no two closed, connected subsets of this space whose union is the punctured disk and whose intersection has a separation of positive distance. However, as we have seen by an example, this punctured disk admits a metric which allows us to consider it as the cylinder $S^1 \times I_1$, $I_1 = [0,1)$.

We now have occasion to use the component decomposition theory we developed in §III. We begin with the following lemma:

Lemma. *Let X be a locally connected, unicoherent metric continuum, and suppose that $f : X \to S^1$ is a mapping and $\mathcal{M}$ is the component decomposition of f with quotient space M. Then M is a dendrite.*

PROOF. By the work of §III, we know that the decomposition function φ of M is closed and monotone, so we see at once that M is a locally connected, unicoherent metric continuum. Also f has the unique factorization $f = l \circ \varphi$, where $l: M \to S^1$ is light. Now assume that M contains a simple closed curve J. Since l is light, we can pick two points of J, say u, v, such that $l(u) \neq l(v)$. Next we choose two points a, b in S^1 which separate $l(u)$ and $l(v)$. Clearly $B = l^{-1}(a) \cup l^{-1}(b)$ must separate M.

Let N_u be the component of $M \sim B$ which contains u, and take N_v to be the component of $M \sim \mathrm{Cl}(N_u)$ which contains v. The boundary of N_v is contained in B, and the set $M \sim N_v$ is closed and connected. We now have $M = (M \sim N_v) \cup \mathrm{Cl}(N_v)$, and since M is unicoherent, $(M \sim N_v) \cap \mathrm{Cl}(N_v)$ is a continuum in B and therefore a single point p. Thus u and v in J are separated in M by the removal of the point p; hence M contains no simple closed curve. $\qquad\square$

3. A locally connected metric continuum X is unicoherent if and only if every mapping $f: X \to S^1$ is exponentially representable.

The sufficiency part was proved in Exercise 2. For the converse we will use a result proved later, that every dendrite has the universal exponential property. Using this fact—as well as the fact just proved that M, the quotient space of the component decomposition $\mathcal{M}$, is a dendrite—we show that X has the universal exponential representation property. Suppose that $f: X \to S^1$ is continuous and $f = l \circ \varphi$ is the factorization mentioned in the last lemma. Let $l: M \to S^1$ be expressed as $e^{iu(m)}$, where $u: M \to R^1$ is a mapping. Then $u \circ \varphi$ is a mapping of X into R^1 such that $f(x) = e^{i(u \circ \varphi(x))}$. Hence the theorem is proved.

4. If X_1 and X_2 are each closed in their union $X = X_1 \cup X_2$, and if $X_1 \cap X_2$ is non-empty and connected, then if $f: X \to S^1$ is exponentially representable on each of X_1 and X_2, f is exponentially representable on X.

Let us assume f has the representations $e^{iu(x)}$ on X_1 and $e^{iv(x)}$ on X_2. Then $u(x) = v(x) + 2k\pi$ on the connected set $X_1 \cap X_2$, so we define

$$w(x) = \begin{cases} u(x) & \text{if } x \in X_1, \\ v(x) + 2k\pi & \text{if } x \in X_2. \end{cases}$$

It is easily seen that $w: X \to R^1$ is well defined. Furthermore, if K is any closed set in R^1, then $w^{-1}(K) = \{u^{-1}(K) \cap X_1\} \cup \{v^{-1}(K - 2k\pi) \cap X_2\}$ is a closed set in X, so w is continuous. Then $e^{iw(x)} = e^{iu(x)}$ on X_1, and $e^{iw(x)} = e^{iv(x)}e^{2\pi ik} = e^{iv(x)}$ on X_2, since $e^{2\pi ik} = 1$, so $f(x) = e^{iw(x)}$ is an exponential representation of f on X.

5. Let $X = \bigcup_{n=1}^{\infty} X_n = \bigcup_{n=1}^{\infty} \mathrm{int}(X_n)$ be a metric space, where $X_1 \subset X_2 \ \cdots$ and each X_n is connected. Then if $f: X \to S^1$ is exponentially representable on each X_n, it is exponentially representable on X.

If X is the union of a finite number of such sets, then the theorem is trivial. Now for each space X_i, suppose f has the exponential representation $f|X_i = e^{iu_i(x)}$. Then, since X_1 is connected, we must have $u_1(x) - u_2(x) = 2k_2\pi$ on X_1, so we define $v_2(x) = u_2(x) + 2k_2\pi$ on X_2, and f has the representation $e^{iv_2(x)}$ on X_2.

Now on X_3 we may get a similar $v_3(x) = u_3(x) + 2k_3\pi$, so that $f(x) = e^{iv_3(x)}$ on X_3.

In general, on X_n we get $v_n(x) = u_n(x) + 2k_n\pi$, and $f(x) = e^{iv_n(x)}$ on X_n is an exponential representation of f, where $v_1 = u_1$.

Define a function $u: X \to \mathbf{R}^1$ by $u(x) = v_n(x)$ for $x \in X_n$; then u is well defined by construction. If W is an open set in $\mathbf{R}^1$, then $u^{-1}(W) = \bigcup_{n=1}^{\infty} (v_n^{-1}(W) \cap \operatorname{int}(X_n))$ is open, so u is continuous. Hence $f: X \to S^1$ has the exponential representation $f(x) = e^{iu(x)}$.

6. A metric space has the universal exponential representation property if for each $\varepsilon > 0$ and each admissible metric, $X = \bigcup_{n=1}^{\infty} D_n$, where each D_n is a closed connected set of diameter $< \varepsilon$ such that $D_{n+1} \cap (\bigcup_{i=1}^{n} D_i)$ is connected, and each x in X is interior to some finite union of these sets D_n. Then X is unicoherent and locally connected.

For an arbitrary uniformly continuous mapping $f: X \to S^1$, we shall show that f is exponentially representable. Choose $\varepsilon > 0$ such that $|f(x_1) - f(x_2)| < 2$ whenever $\rho(x_1, x_2) < \varepsilon$. Then cover X with closed, connected sets of diameter $< \varepsilon$ such that $D_{n+1} \cap (\bigcup_{i=1}^{n} D_i)$ is connected.

The range of D_1 in S^1 is a proper arc, so $f|D_1$ is exponentially representable: $f(x) = e^{iu_1(x)}$. Also $f|D_2$ is exponentially representable, and since $D_1 \cap D_2$ is connected, by Exercise 4 we have an exponential representation $f(x) = e^{iu_2(x)}$ on $X_2 = D_1 \cup D_2$. But $f|D_3$ is exponentially representable, so again we get $f(x) = e^{iu_3(x)}$ on $X_3 = \bigcup_{i=1}^{3} D_i$, since $X_2 \cap D_3$ is connected.

By finite induction, we get $f(x)$ exponentially representable on $X_n = \bigcup_{i=1}^{n} D_i$ with $D_{n+1} \cap X_n$ connected. Now since the D_n's are closed and form a cover for X, we write $X = \bigcup_{n=1}^{\infty} X_n = \bigcup_{n=1}^{\infty} \operatorname{int}(X_n)$, since each $x \in X$ is interior to one of the X_n's. Then the preceding exercise indicates that X has the universal exponential representation property and hence is unicoherent by Exercise 2.

To show local connectedness, let $x \in X$ and let $V_\varepsilon(x)$ be the ε-sphere about x. Cover X with connected sets D_n of diameter $< \varepsilon/2$. Then since $x \in X$ is interior to $\bigcup_{i=1}^{n} D_i = M$, let $K = \bigcup \{D_n \subset M : x \in D_n\}$. Then K is connected and diam $K < \varepsilon$, so that $K \subset V_\varepsilon(x)$. Since $x \in \operatorname{int}(M)$, we have that $x \in \operatorname{int}(K)$ and $\operatorname{int}(K)$ satisfies $x \in \operatorname{int}(K) \subset K \subset V_\varepsilon(x)$, so X is locally connected.

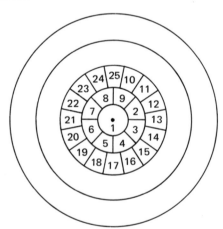

Figure V.2

Remark. We emphasize the importance of the stipulation that the hypothesis be true for each admissible metric. Considering again the punctured disk, we see that it can be covered by closed connected sets of diameter $< \varepsilon$ such that $D_{n+1} \cap (\bigcup_{i=1}^{n} D_i)$ is connected if we follow the procedure in Figure V.2. However, this same space under an equivalent metric becomes the cylinder $S^1 \times [0, \infty)$, which cannot be so covered as to satisfy the hypothesis. Since the punctured disk is multicoherent, the "admissible metric" phrase is seen to be necessary in stating this theorem.

7. I^n, E^n $(n > 0)$, and S^n $(n > 1)$ have the universal representation property and thus are unicoherent.

Take $f : I^n \to S^1$ to be a uniformly continuous mapping which is not exponentially representable. Then divide I^n into closed connected "halves" whose intersection is connected, and call these halves D_1^1 and D_2^1. Then one of these halves is such that $f|D_i^1$ is not exponentially representable, since if not, Exercise 4 of this section implies that $f|I^n$ is exponentially representable. Continue to divide the half of D_j^i on which f is not exponentially representable. Eventually we get a D_j^i of diameter $< \varepsilon$, where ε is such that $\rho(y_1, y_2) < \varepsilon$ in I^n guarantees that $|f(y_1) - f(y_2)| < 2$ by uniform continuity. On this D_j^i, f is exponentially representable, which is a contradiction. Hence f must have been exponentially representable on I^n, so that I^n has the universal representation property and hence is unicoherent.

Next let $E^n = \bigcup_{k=1}^{\infty} (I^n)_k$, being careful to ensure that $(I^n)_{k+1} \cap \bigcup_{m=1}^{k} (I^n)_m$ is connected.

For the sphere S^n, $n > 1$, we write $S^n = S_N^n \cup S_S^n$ where S_N^n and S_S^n are the two hemispheres meeting on S^{n-1}, the equator, which is a connected set. The hemispheres are homeomorphic to the disk bounded by the equator under a polar projection map, and this disk is homeomorphic to I^n, which has the universal exponential representation property. Then the hemispheres also have this property, and an application of Exercise 4 completes the proof that S^n has the property and is therefore unicoherent.

8. Every dendrite $\mathcal{D}$ has the universal exponential representation property.

For some finite subcovering of $\mathcal{D}$ by open connected sets D of diameter $< \varepsilon$, consider the collection $\{\text{Cl}(D)\}$. Select any set and label it D_1; then choose some $\text{Cl}(D)$ which meets D_1 and label it D_2. Evidently $D_1 \cap D_2$ is connected, since $\mathcal{D}$ can contain no simple closed curve. Now this process may be continued, selecting a D_3 which meets $D_1 \cup D_2$ and so on. At each stage $D_{n+1} \cap (\bigcup_{i=1}^{n} D_i)$ is connected or else $\mathcal{D}$ contains a simple closed curve. This process eventually results in the labeling of every set $\text{Cl}(D)$ and $\mathcal{D} = \bigcup_{n=1}^{N} D_n$, and each $x \in \mathcal{D}$ is interior to a finite union of D_n's so Exercise 6 of this section indicates that $\mathcal{D}$ has the universal exponential representation property.

9. If X is a connected, locally connected, unicoherent space, the boundary of each component of the complement of a closed and connected set K in X is itself connected.

Denote the set C as a component of $X \sim K$. Then C is open in X, so ∂C is nonvoid, since X is connected. If E denotes a component of $X \sim K$, then the set $(\bigcup_{E \neq C} \text{Cl}(E)) \cup K$ is connected, since each set E has limit points in the connected set K.

However, no point of C is a limit point of any E, by the maximality of E, so $(\bigcup_{E \neq C} \text{Cl}(E)) \cup K = X \sim C$ is a closed set. Then the unicoherence of X implies that the intersection of $(\bigcup_{E \neq C} \text{Cl}(E)) \cup K$ and $\text{Cl}(C)$, which is ∂C, is a connected set.

10. If the boundary of each component of the complement of a closed, connected set in a locally connected and connected regular space X is itself connected, then X is unicoherent.

We begin by assuming that X can be represented as the union to two closed, connected sets A and B whose intersection has the separation $A \cap B = M \cup N$. Let $A_M = \bigcup C_M$ of components of $A \sim A \cap B$ having boundary points in M, and $A_N = \bigcup C_N$ of components of $A \sim A \cap B$ with boundary points in N. If $p \in N$ is a limit point of A_M, then some component, say C_M, intersects a connected open set U containing p, with

$U \cap M = \varnothing$, and hence C_M has boundary points in N. But this is impossible, since we assumed that ∂C_M was connected. Thus $\mathrm{Cl}(A_M) \cap N = \varnothing$, and similarly $\mathrm{Cl}(A_N) \cap M = \varnothing$. This results in a separation

$$A = (A_M \cup M) \cup (A_N \cup N),$$

since local connectedness and the boundary condition imply that A_M and A_N are disjoint open sets. But this is a contradiction, since A is connected.

Plane Topology

In this last section we specialize to consideration of the Euclidean plane Π and a space $\mathscr{S}^2$ homeomorphic to the sphere S^2.

Definition. A *semipolygon* is a simple closed curve containing a straight line segment.

Definition. A *θ-curve* is a continuum which is the sum of three simple arcs *axb*, *ayb*, *azb* meeting at only their end points, *a* and *b*, which are termed vertices.

Definition. A continuum D is termed a *dygon* provided D is the union $D = A \cup B$ of two continua A and B meeting in two distinct points x and y such that $A - x - y$ and $B - x - y$ are connected. The vertices of D are x and y.

Definition. A unicoherent locally connected continuum is called an $\mathscr{S}^2$ provided that

(a) it contains a simple closed curve;
(b) it is separated by every simple closed curve it contains;
(c) it is not separated by any arc of a simple closed curve.

Alternatively, we can require that $\mathscr{S}^2$ be a space $\mathscr{S}$ such that

(a) $\mathscr{S}$ is a locally connected continuum;
(b) $\mathscr{S}$ contains a simple closed curve;
(c) $\mathscr{S}$ is separated irreducibly by every simple closed curve in $\mathscr{S}$;
(d) $\mathscr{S}$ is unicoherent.

Remark. A topological space $\mathscr{S}$ is an $\mathscr{S}^2$ if and only if it is homeomorphic to the sphere S^2. This is a result which will not be proved in this text.

Definition. A *nonseparating mapping* $f: X \Rightarrow Y$ is a mapping with the property that $f^{-1}(y)$ does not separate X for any point $y \in Y$.

EXERCISES VI.

1. *Jordan Curve Theorem*: Any simple closed curve J in the plane Π separates Π into exactly two regions and is the boundary of each region.

2. *Plane Separation Theorem*: If A is a continuum and B is a closed connected set in Π or $\mathscr{S}^2$ with $A \cap B = T$, a totally disconnected set, and with $A - T$ and $B - T$ connected, then there exists a simple closed curve J in Π separating $A - T$ and $B - T$, such that $J \cap (A \cup B) \subset A \cap B = T$.

3. The property of being an $\mathscr{S}^2$ is invariant under a nonconstant, monotone, non-separating mapping.

SOLUTIONS.

Theorem. *If a closed curve C in S^2 separates S^2, then there exists a mapping $f : C \to S^1$ which is not homotopic to a constant.*

PROOF. We suppose that $S^2 - C$ yields a separation $R \cup S$ where R is contained in the northern hemisphere N, with north pole p. Let q be the south pole and $r : S^2 - p - q \to S^1$ be the meridianal projection of $S^2 - p - q$ onto the equator S^1. We assert that $r : C \to S^1$ is not homotopic to a constant. Assuming that r is homotopic to a constant, the homotopy extension theorem says that $r|C$ has a continuous extension to all of S^2, say $g : S^2 \to S^1$. Define G by

$$G(s) = \begin{cases} r(x) & \text{if } x \in B - q, \\ g(x) & \text{if } x \in A, \end{cases}$$

where $R \cup C = A, S \cup C = B$. Thus G is a mapping defined on N, a region in S^2 bounded by S^1 together with S^1. But this is impossible, since $G|S^1$ is the identity on S^1, contradicting Exercises 6 and 7 of §IV, which say that the identity map on S^1 cannot have an extension to the region bounded by S^1, since the identity mapping on S^1 cannot be homotopic to a constant. ☐

Corollary. *No simple arc or dendrite separates the plane.*

PROOF. Using the standard homeomorphism (Exercise 5, §IV) between the punctured sphere $S^2 - p$ and the plane Π, if an arc or dendrite were to separate Π, then an arc or dendrite must separate S^2. If D is such an arc or dendrite, then there exists a mapping $f : D \to S^1$ which is not homotopic to a constant. However, D has the universal exponential representation property, so if $f(x) = e^{iu(x)}$, then $h(x,t) = e^{itu(x)}$ is a homotopy from f to a constant map, which is a contradiction. Therefore D cannot separate Π. ☐

Lemma. *Every component of $\Pi - J$ is bounded by all of J.*

PROOF. If U is a component of $\Pi - J$, then Exercise 9, §V says that ∂U is connected. But ∂U separates Π into the sets U and $X - \bar{U}$. Since $\partial U \subset J$, if $\partial U \neq J$, then a simple arc divides the plane Π, which contradicts the preceding corollary. ☐

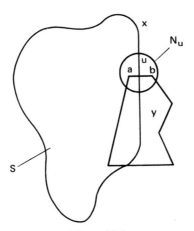

Figure VI.1

1. *Jordan Curve Theorem*: Any simple closed curve J in the plane Π separates Π into exactly two regions and is the boundary of each region.

The second part of this assertion has been proved above; for the rest, we assume that a polygon divides the plane and show that a semipolygon does also.

Let S be a semipolygon containing the segment $[x,y]$, and u be a point of the segment with neighborhood N_u meeting S only on the segment $[x,y]$ (see Figure VI.1). Now choose points a, b in distinct "halves" of N_u; if no points in N_u are separated by S, then join a to b with a polygonal arc not meeting S. Then we connect a to b with a straight line segment lying in N_u and call this segment, together with the polygonal arc above, P, a polygon. Then x and y must be separated by P, but this is not true if S does not meet the polygonal arc connecting a to b, since S is not separated by a single point. The contradiction thus asserts that S divides Π.

Hence there exist at least two components of $\Pi - S$. Suppose there are three; let a, b, and c be points in each of these components. Since J is the boundary of each component, we can find an open set V about u and arcs $[aa_1]$, $[bb_1]$, and $[cc_1]$ in the respective components such that $a_1, b_1, c_1 \in V$. But then if x' and y' are the points of $[x,y]$ on ∂V, then $\partial V - x' - y'$ has three components, which is a contradiction.

A simple closed curve J can be represented as the union of two semipolygons minus a segment by taking distinct points x, a, y, and b in J and connecting a to b with a segment as shown in Figure VI.2. Define $J_x = [axb] + [ab]$ and $J_y = [ayb] + [ab]$, two semipolygons. Then let $\Pi - J_x = R_x + D_x$ and $\Pi - J_y = R_y + D_y$, where D_x contains $R_y \cup [ayb]$ and D_y contains $R_x \cup [axb]$.

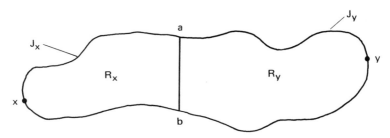

Figure VI.2

Then $\Pi - J = D_x \cap D_y + R_x + R_y + (a,b)$. Now $D_x \cap D_y$ does not have a limit point on (a,b), for if $q \in (a,b)$ is a limit point of $D_x \cap D_y$, we can construct a segment $(q,p]$ lying in $D_x \cap D_y$ and a circle B about q missing J and meeting $(q,p]$ at some point other than p. If this circle meets (a,b) at u and v, then $B - u - v$ has components in R_x, R_y, and $D_x \cap D_y$, which is impossible, since $B - u - v$ has only two components. Hence the set $D_x \cap D_y$ is open and closed in $\Pi - J$, so J separates Π.

To show that $\Pi - J$ has only two components, assume that R, S, and T are components of $\Pi - J$. Then, since J is the boundary of each component, we may pick a, b, c, and d on J, join a to b with a simple arc A_1 such that $A_1 - a - b \subset R$, and join c to d with a simple arc A_2 such that $A_2 - c - d \subset S$. (See Figure VI.3.) If by $[da]$ and $[bc]$ we mean the simple arcs in J with these prescribed end points, and meeting no other of the points a, b, c, d, then the set $J' = [da] + A_1 + [bc] + A_2$ is a simple closed curve.

Now $T \cap J' = \varnothing$, so that T lies in a component V of $\Pi - J'$, and $\Pi - J'$ contains at least two components U and V. Since $\partial T = J$, then V contains $J - J'$, so $U \cap J = \varnothing$. However, $\partial U = J'$ has limit points on A_1 and A_2, and therefore U meets R and S,

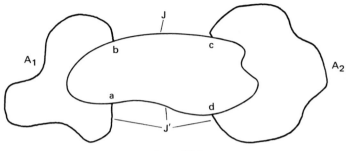

Figure VI.3

since $A_1 - a - b \subset R$ and $A_2 - c - d \subset S$. Hence U must meet J; but $U \cap J = \varnothing$, so a contradiction arises.

θ-curve Theorem *A θ-curve separates the plane into exactly three regions, each bounded by a unique pair of the arcs whose union is θ.*

PROOF. Let $\theta = [axb] + [ayb] + [azb]$, three arcs which meet only at a and b, and define $[ayb] + [azb] = J_x, [axb] + [azb] = J_y,$ and $[ayb] + [axb] = J_z$. By the Jordan Curve Theorem, we write $\Pi - J_x = R_x \cup D_x, \Pi - J_y = R_y \cup D_y,$ and $\Pi - J_z = R_z \cup D_z$, where we understand that $x \in D_x$, $y \in E_y$, and $z \in D_z$. Thus $\Pi - \theta = R_x \cup R_y \cup R_z \cup (D_x \cap D_y \cap D_z)$, where each region R_x, R_y, R_z is a component and is bounded by a unique pair of arcs. We must show that $D_x \cap D_y \cap D_z$ is empty. Assume not, and let U be a component of $D_x \cap D_y \cap D_z$. Now, since U is in D_z, U has boundary points on J_z, so we choose p and q on J_z and join them with a simple arc $[pq]$ lying in $U + p + q$. We also join two other points r and s in J_z with an arc lying in $R_z + r + s$. (See Figure VI.4.)

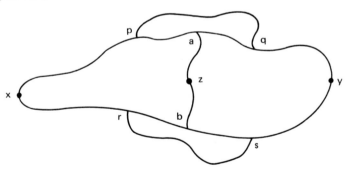

Figure VI.4

Now we have a simple closed curve $J = [pq] + [qys] + [sr] + [rxp]$ which separates Π into the regions A and B.

The local connectedness of U ensures that for any point t of $[pq]$ in U, there is an open neighborhood N_t about t, lying entirely in U. But $t \in \partial A$ and $t \in \partial B$, so the neighborhood N_t must meet A and B, so U meets A and B. Since an argument similar to the above shows that A and B also meet R_z, the connectedness of A and B implies that A and B meet J_z on $\theta - J$. However, the set $J_z \cap (\theta - J) = [paq] + [sbr]$ is connected by the arc $[azb]$, which also lies in $(\Pi - J) \cap J_z$. The two components of

$\Pi - J$ then meet the same connected set in $\Pi - J$, which is a contradiction and establishes the theorem.   □

The next major result we want to obtain is the Plane Separation Theorem. To prove this we first must establish some preliminary results, beginning with the following lemma.

Lemma 1. *If the continuum M is not locally connected at one of its points p, there exists a spherical neighborhood R with center p and an infinite sequence of distinct components $N_1, N_2, \ldots$ of $M \cap \mathrm{Cl}(R)$ converging to a limit continuum N which contains p and has no point in common with any of the continua $N_1, N_2, \ldots .$*

PROOF. If local connectedness fails at $p \in M$, then for some spherical neighborhood R of radius $\varepsilon > 0$ about p, for every positive δ, $V_\delta(p)$ contains points of $M \cap \mathrm{Cl}(R)$ which do not lie in the component C of $M \cap \mathrm{Cl}(R)$ containing p.

Choose x_1 in $V_{\varepsilon/2}(p)$ not in C and let C_1 be the component of $M \cap \mathrm{Cl}(R)$ containing x_1. Then C_1 is closed and does not contain p, so there exists a point x_2 in $V_{\varepsilon/4}(p)$ and not in $C_1 \cup C$. For suppose the contrary. Then by the observation first stated in the proof, p must be a limit point of C_1, so that $p \in C_1$, which is a contradiction. Denote the component of $M \cap \mathrm{Cl}(R)$ containing x_2 as C_2. Similarly $C_1 \cup C_2$ is closed, so there exists a point x_3 in $V_{\varepsilon/8}(p)$ not in $C_1 \cup C_2 \cup C$, and we denote the component of $M \cap \mathrm{Cl}(R)$ containing x_3 as C_3.

Continuing this process indefinitely, we get a sequence $C_1, C_2, \ldots$ of distinct continua [with $x_i \in C_i$ and $\{x_i\} \cap (\bigcup_{k=1}^{i-1} C_k) = \varnothing$] whose limit inferior contains p. By Exercise 2, Part A, §VIII, $[C_i]$ has a convergent subsequence, which we may term (N_i), having a limit N, itself a continuum by Exercise 1, Part A, §XI. Clearly N must contain p, and thus N is contained in C, so $N \cap \bigcup_{i=1}^{\infty} N_i = \varnothing$.   □

Corollary. *A continuum M which is not locally connected at a point p necessarily fails to be locally connected at all points of a nondegenerate subcontinuum of M.*

PROOF. If, in the proof of Lemma 1, ε is the radius of R and N is the limit continuum, let H be the component of $N \cap \mathrm{Cl}(V_{\varepsilon/2}(p))$ containing p. Then M cannot be locally connected at any point of H and H must be non-degenerate since it contains p and at least one point of $\partial V_{\varepsilon/2}(p)$ by Exercise 1, Part A, §XII.   □

Torhorst Theorem. *The boundary B of each component C of the complement of a locally connected continuum M is itself a locally connected continuum.*

PROOF. Both Π and $\mathscr{S}^2$ are unicoherent, so Exercise 9, §V (the Phragmen–Brouwer Theorem) indicates that B is connected, and since $B \subset M$, B is compact. Let local connectedness of B fail at p in B. Then we choose an $\varepsilon > 0$ such that for no positive δ do all the points of $V_\delta(p) \cap B$ lie in a component of $M \cap \mathrm{Cl}(R)$ containing p, where R is the spherical neighborhood of radius ε about p. Using the construction in the lemma, we see that $\delta(N_i) > \varepsilon/2$ for all i, since each component N_i must meet the boundary of R and $x_i \in N_i$ where $\rho(x_i, p) < \varepsilon/2$. Note that $N_i \subset B$ for each i. Now because M has property S, we can cover M with a finite union of locally connected continua $A_i \subset M$ of diameter $< \varepsilon/10$.

Since there are an infinite number of distinct components N_i and a finite number of A_i, we can choose locally connected continua A_1 and A_2 from this cover with $\rho(A_1,A_2) > \varepsilon/4$, and A_1, A_2 meet an infinite number of the same sets N_i. We need only three such sets, which we may designate as K_1, K_2, and K_3 (see Figure VI.5). The K_i are disjoint compact sets, so if $\sigma = \min \rho(K_i,K_j) \cdot (\frac{1}{10})$, where $i \neq j$ and $i,j = 1, 2, 3$, then $\sigma > 0$. Regions R_i about K_i may be chosen to lie in σ-neighborhoods of K_i, that is, $R_i \subset V_\sigma(K_i)$.

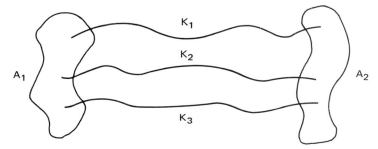

Figure VI.5

These regions R_i are arcwise connected, so we may choose simple arcs L_i in R_i connecting p_i to q_i, where $p_i \in A_1 \cap R_i$ and $q_i \in A_2 \cap R_i$. Then we redefine p_i, q_i to be the first points in A_1, A_2 respectively, met by L_i (see Figure VI.6).

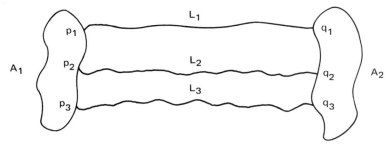

Figure VI.6

Now A_1 and A_2 are arcwise connected, so we join p_1 to p_2 with a simple arc in A_1 and then join p_3 to p_2 with a simple arc in A_1, taking a to be the first point of arc $[p_1 p_2]$ met by the arc $[p_2 p_3]$. A similar procedure in A_2 obtains the arcs $[q_1 q_2]$ and $[q_2 q_3]$, with $[q_1 q_2]$ first meeting $[q_2 q_3]$ at b. The set $\theta = L_1 + L_2 + L_3 + p_1 a + p_2 p_3 + q_1 b + q_2 q_3$ is a θ-curve with vertices a and b (see Figure VI.7). This curve $\theta \subset M$, and hence

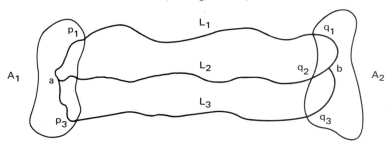

Figure VI.7

$C \cap \theta = \varnothing$. (Recall that C is the component of the complement of M whose boundary is B.) Therefore C must lie entirely in a single component of the complement of θ.

However, let $p \in L_i$ and observe that $V_{2\sigma}(p)$ must meet $K_i \subset B$ and hence contain a point of C. Hence there are points of C in more than one component of the complement of θ, which establishes theorem.   □

Lemma 2. *The boundary B of each component R of the complement of M, a locally connected continuum having no cut point, is a simple closed curve.*

PROOF. Since B separates Π or $\mathscr{S}^2$, B must contain a simple closed curve J. If $B - J \neq \varnothing$, then choose $p \in B - J$ and connect p to J with a simple arc in B which first meets J at q. This construction is possible because B is a locally connected continuum and hence arcwise connected. Now since p and J lie in M, we can join p to J with a simple arc which first meets J at $a \neq q$ such that $[pq] \cap [pa] = p$. This is possible because M contains no cut point. (See Figure VI.8.) Then the set $[pq] + [pa] + J$ is a θ-curve with vertices at a and q, and $\theta \cap R = \varnothing$, since $\theta \subset M$. However, p is a boundary point of R and $J \subset \partial R$, so the region R, which must lie in a component of the complement of θ, has a boundary point on each arc of θ, which is a contradiction.   □

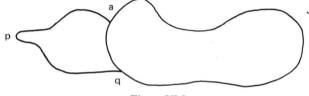

Figure VI.8

Lemma 3. *No region in a locally connected unicoherent space X has a cut point, if X has no cut point.*

PROOF. Assume X is locally connected and unicoherent and has no cut point, but that M is a region in X with a cut point p. Let R be a component of $M - p$. Then R is open. Let $S = \mathrm{Cl}(R) \cup \{C : C$ is a component of $X - M$ containing a limit point of $R\}$ and $T = \mathrm{Cl}(M - R) \cup \{D : D$ is a component of $X - M$ containing a limit point of $M - R\}$. Then $\mathrm{Cl}(S)$ and $\mathrm{Cl}(T)$ are each nonempty, closed, and connected, and $\mathrm{Cl}(S) \cup \mathrm{Cl}(T) = X$. However, p is the only point of M in $\mathrm{Cl}(S) \cap \mathrm{Cl}(T)$. This follows because M is open, so by the construction of S only points of $\mathrm{Cl}(R)$ are in $\mathrm{Cl}(S) \cap M$; i.e., $\mathrm{Cl}(S) \cap M \subset \mathrm{Cl}(R)$. Also, since R is open, $\mathrm{Cl}(T) \cap M$ meets $\mathrm{Cl}(R)$ only at the point p. Therefore $\mathrm{Cl}(S) \cap \mathrm{Cl}(T) \cap M = p$.

Clearly p is not all of $\mathrm{Cl}(S) \cap \mathrm{Cl}(T)$, since X has no cut point, so since p is open and closed in $\mathrm{Cl}(S) \cap \mathrm{Cl}(T)$, then $\mathrm{Cl}(S) \cap \mathrm{Cl}(T)$ is not connected, contradicting unicoherence.   □

2. *Plane Separation Theorem*: If A is a continuum and B is a closed connected set in Π or $\mathscr{S}^2$ with $A \cap B = T$, a totally disconnected set, and with $A - T$ and $B - T$ connected, then there exists a simple closed curve J in Π separating $A - T$ and $B - T$, such that $J \cap (A \cup B) \subset A \cap B = T$.

Let $H_1 = \{x \in A : \rho(x,B) \geq 1\}$, and for $n > 1$ define $H_n = \{x \in A : 1/n \leq \rho(x,B) \leq 1/n - 1\}$. Then each set H_n is compact, and since Π has property S locally, each set H_n may be covered by a finite number of regions of diameter $< 1/n$. If we call such regions

$R_i^{(n)}$ and say $K_n = \bigcup_{i=1}^{N_n} R_i^{(n)}$, then we claim that $K = \text{Cl}(\bigcup_{n>0} K_n)$ is a locally connected continuum. Local connectedness cannot fail on $K - T$, since any point $p \in K - T$ is a point or a limit point of only a finite number of regions $R_i^{(n)}$ which contain a δ-neighborhood of p which is connected. Local connectedness holds also for T, since if not, then by the corollary to Lemma 1, K contains a nondegenerate continuum, at each point of which K is not locally connected. Then T must contain a nondegenerate continuum, which is a contradiction, since T is totally disconnected. That K is compact follows, since it is a closed and bounded set in Π, or a closed subset of $\mathscr{S}^2$. Now the interior of K is the union of regions which each meet $A - T$ and whose union includes $A - T$, a connected set; therefore $\text{int}(K)$ is itself connected. It follows that K is connected.

We now claim that K has no cut point, for if p is interior to K, then p lies in some region and is not a cut point, by Lemma 3. However, $\text{int}(K)$, being the union of regions which meet the connected set $A - T$, is itself connected, so if $p \in K - \text{int}(K)$, then $K - p$ is connected, since $\text{int}(K) \subset K - p \subset \text{Cl}(\text{int}(K)) = K$. Hence K contains no cut point, and we may apply Lemma 2 to the component U of the complement of K containing $B - T$. That $B - T \subset U$ follows, since $(B - T) \cap K = \varnothing$ by construction of K and $B - T$ is connected. Thus $\partial U = J$, a simple closed curve, and since $A - T$ is interior to K and $B - T$ is interior to U, we see that $J \cap (A \cup B) \subset T$.

Before stating the Dygon Theorem, we define the *edges* of a dygon D to be the pair of connected sets resultant after removing the vertices from D.

Dygon Theorem. *Every dygon D separates Π or $\mathscr{S}^2$, and there are two components of the complement of D which have the vertices x and y on their boundaries. No other component of the complement of D has a boundary meeting both edges of D.*

PROOF. Let D be a dygon with edges E_1 and E_2 and vertices x and y. First we will show that the complement of D is separated. We assume otherwise, and first construct by the Plane Separation Theorem a simple closed curve J which meets D only at x and y and separates E_1 and E_2. We choose points a, b in J such that $a \cup b$ separates x and y in J, and join a to b with a simple arc which misses D. Then define α to be the last point in $[ab]$ which is in $[xay]$ and β to be the first point in $[\alpha b]$ which is in $[xby]$, to obtain a simple arc $[\alpha\beta]$ which misses $J - \alpha - \beta$, but such that $\alpha \cup \beta$ still divides x and y (see Figure VI.9). Then $J \cup [\alpha\beta] = \theta$ is a θ-curve with vertices at α and β, and hence its complement has three components. However, E_1 and E_2 must lie in the same component, since they both have x and y as limit points; but this contradicts the Plane Separation Theorem.

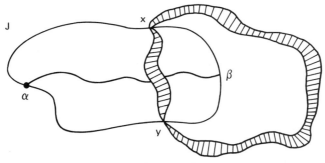

Figure VI.9

Next we show that there are two components U and V of the complement of D such that $x, y \in \partial U \cap \partial V$. Let U be the component containing, α and V the one containing β. Then the open arc $[x\alpha y]$ is connected in the complement of D, and hence $[x\alpha y] \subset U$, so that ∂U contains x and y. Similarly, the open arc $[x\beta y]$ is connected in the complement of D, and hence $[x\beta y] \subset V$ so that ∂V contains x and y.

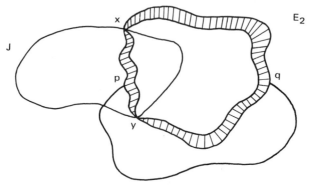

Figure VI.10

Now suppose a third component W of the complement of D also has a boundary ∂W which meets E_1 and E_2. Then we can construct a simple arc from E_1 to E_2, say $[pq]$, such that $[pq] - p - q$ lies wholly in W (see Figure VI.10). By the Plane Separation Theorem, this open arc $[pq]$ must meet J in some point r in the complement of D. Then, because r is in U or in V, and in W, we have that W meets either U or V, which is impossible. Hence ∂W cannot meet E_1 and E_2. □

Corollary. *Let $\varphi : \mathscr{S} \Rightarrow \Sigma$ be a monotone mapping, where $\mathscr{S}$ is an $\mathscr{S}^2$. Any simple closed curve J in Σ containing distinct points x and y with unique φ-inverses separates Σ. If the boundary of a component of $\Sigma - J$ meets each open arc xy of J, then the boundary must contain x and y.*

PROOF. We use the fact that $\varphi^{-1}(J)$ is a dygon D in $\mathscr{S}$. For, if two arcs of J are $[xay]$ and $[aby]$ in Σ, then $\varphi^{-1}(J)$ is a compact connected set which is separated by the removal of $u = \varphi^{-1}(x)$ and $v = \varphi^{-1}(y)$. Let $E_1 = \varphi^{-1}([xay]) - u - v$ and $E_2 = \varphi^{-1}([xby]) - u - v$ be the edges of $\varphi^{-1}(J)$. Then E_1 and E_2 are each connected, since φ is closed and monotone. Since D is an inverse set which separates $\mathscr{S}$, $\varphi(D) = J$ separates Σ by the closedness and monotonicity of φ.

If U is a component of $\Sigma - J$ whose boundary meets both arcs of $J - x - y$, then $\varphi^{-1}(U)$ is a component of $\mathscr{S} - D$ whose boundary meets E_1 and E_2, so $\partial \varphi^{-1}(U)$ contains u and v. Therefore $\varphi(\partial \varphi^{-1}(U)) = \partial U$ contains x and y. □

3. The property of being an $\mathscr{S}^2$ is invariant under a nonconstant, monotone, nonseparating mapping.

Let $f : S \Rightarrow \Sigma$ be such a mapping, where S is an $\mathscr{S}^2$. It is easily checked that Σ is a locally connected unicoherent continuum, since f is monotone and closed, and Σ has no cut point, since f is nonseparating; so Σ contains a simple closed curve J. If Σ is not an $\mathscr{S}^2$, then either (1) $\Sigma - J$ is connected or (2) $\Sigma - J$ contains a component whose boundary is a proper arc $[ab]$ of J. Let x and y separate a and b in J in case (2) holds.

First of all we define a mapping $\varphi : S \to S'$ as follows: φ is the identity on $S - f^{-1}(x) - f^{-1}(y)$, and φ maps $f^{-1}(x)$ and $f^{-1}(y)$ into these sets each considered as a point of S'. The space S' looks like individual points of $S - f^{-1}(x) - f^{-1}(y)$ together with the "points" $f^{-1}(x) = p$ and $f^{-1}(y) = q$. Clearly φ is a nonconstant, monotone, non-separating mapping, so S' is a nondegenerate, unicoherent, locally connected continuum containing a simple closed curve J':

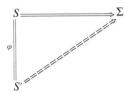

Let c and d be distinct points of J', and since every point of S' has a unique φ-inverse except for p and q, choose u and v in J' to separate c and d, where u, v have unique φ-inverses. Applying the corollary, we see that the arc $[cd]$ cannot be the boundary of a component of $S' - J'$, since such a boundary would meet both arcs of $J' - u - v$ and hence contain u and v. Thus no simple arc of J' separates S'.

Now form the mapping $g : S' \Rightarrow \Sigma$ by $g(s) = f \circ \varphi^{-1}(s)$. It is quickly verified that g is monotone, nonseparating, and nonconstant. The points x, y of J in Σ have unique g-inverses, so by the corollary (and a repetition of the above argument), J separates Σ and no component of $\Sigma - J$ has a boundary consisting of the simple arc ab. Hence Σ is an $\mathscr{S}^2$.

Appendix
Dynamic Topology

G.T. Whyburn, University of Virginia

1. Introduction. This title is not meant to be either laudatory or derogatory but simply to be in contrast with what might be called static topology. The paper of mine which attracted the notice of the Chauvenet Prize Committee in the 1930's was called "On the structure of continua" and dealt largely with cyclic element [1] and other types of structure analysis of certain topological spaces. Even in this, however, the problem was formulated as to what type of mappings will preserve *all* topological properties, especially when applied to special domain spaces such as Euclidean spaces.

Briefly put, dynamic topology refers to the body of results of a topological nature in which the function concept plays an essential role. The function concept is basic in nearly all fields of mathematics. We study spaces and systems by observing how they act when undergoing transformations affected by the action of functions upon them. This is similar to methods used in the physical and natural sciences in studying concrete problems and phenomena of the world around us. In most cases movement, growth, change under natural or artificially induced conditions, evolution, or the like appropriate to the medium under review are involved in physical, chemical or biological studies. In mathematics the function concept enjoys an ever increasing role in nearly all areas, whether it be analysis, algebra or topology. This is especially true now and is of fairly recent vintage in topology. However, there is no lessening of interest and activity in the static study of structures. The Poincaré conjecture, for example, is the source of much fruitful effort as are also manifold structure and characterization and the relation between a set and its complement in Euclidean and other manifolds [2, 3, 4]. Even these areas of topology, however, become more and more pervaded by usage of the function device. How a space changes or remains invariant under action of various types of functions is most revealing as to its structure.

Only a few phases of this subject will be dealt with in this lecture and these are selected more because of their relevance to my own work and interests than on other grounds. However, it is hoped that they will be of interest to you also, at least in that they are significant examples of the work going on in certain areas of topological research. We will be concerned largely with results and problems related to compactness of mappings and compactifications of mappings. Thus an appropriate subtitle would be "Compactness and compactifications of mappings."

To simplify our base of discussion, unless otherwise indicated, all topological spaces will be assumed to be Hausdorff spaces, i.e., distinct points lie in disjoint open sets. A *mapping* $f: X \to Y$ of a space X into another space Y is a *continuous* function from X, its *domain*, to Y, its *range*. The double arrow $f: X \Rightarrow Y$ will indicate that the function maps X *onto* Y. The boundary $\overline{U} - U$ of an open set U is denoted by ∂U.

Reprinted from the *American Mathematical Monthly*, Vol. 77, No. 6, June-July 1970, pp. 556-570.

2. Compactness and quasi-compactness. A mapping $f: X \to Y$ is *compact* [5, 6] provided the inverse $f^{-1}(A)$ of every compact set A in Y is compact. Since X and Y are Hausdorff spaces, note that all mappings $f: X \to Y$ are compact when X is a compact space. However, the mapping $w = e^{i\theta}$ of $0 \leqq \theta < 2\pi$ onto $|w| = 1$ is not compact, for example. In general a mapping is compact if it is closed and has compact point inverses. The converse holds precisely [7] when X and Y are Hausdorff spaces and Y is a k-space, i.e., when the topology in Y is determined by its compact subsets in the sense that a set in Y is closed if and only if it meets every compact set in a closed set. Assuming the above conditions, compactness of a mapping also is readily characterized in terms of filters or directed families, $f: X \Rightarrow Y$ being compact if and only if for every directed family N in Y converging to $y \in Y$ the inverse family $f^{-1}(N)$ is directed toward $f^{-1}(y)$ in the sense that every directed underfamily has a cluster point in $f^{-1}(y)$. Thus compactness of a mapping means precisely that compactness of sets is invariant both forward and backward, under both f and f^{-1}. It is such a useful property because it assures one of action on noncompact domain spaces closely resembling the action of a mapping on compact spaces with which so much can be done.

A mapping $f: X \to Y$ is *quasi-compact* [8] provided the image of every closed (open) inverse set is closed (open). This property was introduced by Alexandroff-Hopf [9] and called strong-continuity. More recently it has come to be known as the property of being a quotient mapping. It will be noted at once that every closed mapping is automatically quasi-compact, as is also every open mapping, and in particular every compact mapping, granted in the latter case that the domain and range spaces are Hausdorff and that the range is a k-space. It also readily follows that every retraction is quasi-compact. In particular note that the exponential mapping $w = e^z$ is quasi-compact but not compact. The same is true of the mapping given by any transcendental entire function.

Quasi-compactness is closely related to and is especially useful in the study of mappings generated by decompositions of a given space X. Indeed if X is decomposed into a collection G of disjoint sets $\{g\}$ and a new space Y' is set up with elements of G as its points and a topology introduced by defining a set in Y' to be open if and only if the union of its elements is an open set in X, the so-called quotient topology, then the natural mapping $\phi: X \Rightarrow Y'$ which maps each $x \in X$ into the element g of G containing x is quasi-compact. If we begin with a mapping $f: X \Rightarrow Y$ and let X be decomposed into the collection $\{f^{-1}(y)\}$ of point inverses, $y \in Y$, then let $\phi: X \Rightarrow Y'$ be the natural mapping of this decomposition, the function $h: Y' \Rightarrow Y$ defined by $h(y') = f\phi^{-1}(y')$ is 1-1 and continuous. Further h^{-1} will be continuous and thus h will be a homeomorphism if and only if f is quasi-compact.

Openness and closedness of a mapping are nicely characterized in terms of quasi-compactness as follows: A mapping $f: X \Rightarrow Y$ is open (closed) if and only if it is quasi-compact and generates a lower (upper) semi-continuous decomposition of X into point inverses. Another remarkable property of quasi-compact mappings is that local connectedness is invariant under all such mappings, thus under all open or closed mappings and all retractions. The latter results from the unexpected fact that a mapping $f: X \Rightarrow Y$ quasi-compact on any cross section (i.e., any set in X mapping onto Y) is automatically quasi-compact on X.

The inverse does not hold, however. Indeed, in contrast to the situation with open or closed mappings, it does not follow [10] that a quasi-compact mapping is quasi-compact when restricted to an inverse set—not even to its kernel [= the set of all $x \in X$ for which $x = f^{-1}f(x)$]. This fact has important implications in the case of invariance of connectedness under the inverse f^{-1} of a mapping $f: X \rightarrow Y$. In case point inverses are connected then X will be connected if Y is connected, provided f is quasi-compact. However, it does not follow from this that the restriction $f | I$ of f to an inverse set I will enjoy the same property, because this restriction may fail to be quasi-compact. What is needed here is that f be hereditarily quasi-compact, i.e., quasi-compact on every inverse set. It turns out that this latter property can be assured for *all* quasi-compact mappings onto a given range space Y, provided Y is an *accessibility space* [11]. A regular Hausdorff space Y is an accessibility space provided each limit point p of a set M in Y is *accessible by closed sets* in Y in the sense that $M + p$ contains a closed set C having p as a limit point. Any quasi-compact mapping onto a regular accessibility space is automatically hereditarily quasi-compact. Thus, in particular, if point inverses for such a mapping are connected then connectedness is fully invariant under the inverse. The notion of accessibility space is readily extended to spaces other than regular. Thus it may be said that for mappings with connected point inverses, connectedness is invariant both forward and backward, provided the mapping is hereditarily quasi-compact and this is true for all quasi-compact mappings into accessibility range spaces. In particular it holds for such mappings into locally compact or first countable range spaces as these are always accessibility spaces. We may summarize by saying: under a compact mapping compactness is invariant both forward and backward; and under a quasi-compact weakly monotone mapping onto an accessibility space, connectedness is invariant both forward and backward.

3. Monotoneity. We next consider compactness and quasi-compactness of a mapping and the compact part of a mapping in a somewhat more restricted setting involving local compactness of the domain and monotoneity of the mapping. A mapping $f: X \rightarrow Y$ is *monotone*, provided its point inverses $f^{-1}(y)$ are continua, i.e., compact connected sets. The sequence of results which we indicate next leads to interesting applications for mappings on the Euclidean spaces.

(1) *If X is locally compact, every quasi-compact monotone mapping $f: X \Rightarrow Y$ is closed (and thus compact) and its range Y is also locally compact.*

To prove this, take any closed set K in X and any $p \in Y - f(K)$, and let $P = f^{-1}(p)$. Then take an open set U about P with $\overline{U}$ compact and $U \cdot K = \emptyset$, using compactness of P and the fact that X is a Hausdorff space. Then ∂U and $C = f(\partial U)$ are compact and thus closed. Let $W = Y - C$ and $R = U \cdot f^{-1}(W)$. Then

R is open, $P \subset R \subset U$, and $R = f^{-1}f(R)$ because $f^{-1}(y)$ is connected and meets R for each $y \in f(R)$ but cannot meet ∂U. Accordingly $f(R)$ is open and contains p but does not meet $f(K)$. Thus $f(K)$ is closed. That Y is locally compact follows from the invariance of local compactness under closed mappings.

(2) *If X and Y are noncompact but locally compact spaces, a mapping $f: X \Rightarrow Y$ is compact if and only if it has a continuous extension f^* from the 1-point compactification X^* of X to the 1-point compactification Y^* of Y.*

The extension is unique when it exists. To prove (2) we suppose f is compact. Let $X^* - X = x^*$, $Y^* - Y = y^*$ and define $f(x^*) = y^*$. Let V be any open set in Y^*. If V contains y^*, $Y^* - V = Y - V$ and this set is compact. Accordingly

$$f^{-1}(Y - V) = X^* - f^{-1}(V) = X - f^{-1}(V)$$

and this set is compact since $f | X$ is compact. Hence $X^* - f^{-1}(V)$ is closed so that $f^{-1}(V)$ is open in X^*. If V doesn't contain y^*, $f^{-1}(V)$ is open in X and thus also in X^*. Hence the extension of f to X^* is continuous.

On the other hand, suppose f has a continuous extension $g: X^* \to Y^*$, so that $f = g | X$. Then since X^* and Y^* are compact but Y is noncompact, g must be an onto mapping and it obviously is compact. Thus f is compact, because X is an inverse set for g and any restriction of a compact mapping to an inverse set is compact.

(2′) COROLLARY. *If X and Y are locally compact, a monotone mapping $f: X \Rightarrow Y$ is compact if and only if it has a continuous monotone extension from X^* to Y^*.*

Proof for Corollary (2′). In case neither X nor Y is compact the proof for (2) shows that when f is compact its extension to X^* is monotone when f is monotone, and that when f has a continuous extension to X^*, f is automatically compact.

There remains the case in which X or Y is compact. If X is compact, Y is necessarily compact and all mappings from X to Y are compact and all have trivial extension to X^*, which is the same as X. If Y is compact, then for $f: X \Rightarrow Y$ compact, $X = f^{-1}(Y)$ is compact, and again f is trivially extendable to X^*. On the other hand, still assuming Y compact, if f has a monotone extension g to X^* and if X^* were different from X, say $X^* - X = x^*$, we would have $g(x^*) = y_0 \in Y$ and $g^{-1}(y_0) = x^* + f^{-1}(y_0)$, a disconnected set, contrary to monotoneity of g.

Monotoneity is essential in (2′). For if f is the 1-1 mapping of a plane (or punctured sphere) obtained by identifying the two poles of a sphere and then deleting one of the poles from the domain of f, then f has a continuous but non-monotone extension to its 1-point compactification (the sphere) but f is not compact of course.

(2″) COROLLARY. *Every rational function $w = f(z)$ generates a compact mapping of the z-plane onto the w-plane.*

For any function $f: X \Rightarrow Y$ and any subset Y' of Y, any subset X' of X

mapping *onto Y'* is called a *trace* of Y' under f. Note that a trace of Y itself is the same as a *cross section* of f as used earlier in this discussion..

(3) THEOREM. *If* $f:X \Rightarrow Y$ *is a monotone mapping where* X *is locally compact, then any set* H *in* Y *which has a compact trace* K *has a compact inverse* [12].

We set $N = f^{-1}(H)$ and consider the restriction $f|N = g$ of f to N. Since K is compact the mapping $g|K:K \Rightarrow H$ is compact and thus, in particular, quasi-compact. Since K is a cross section for the mapping $g: N \Rightarrow H$, it follows that g is quasi-compact. Since g is also monotone it follows by (1) that g is compact. Accordingly the set $g^{-1}(H) = N = f^{-1}(H)$ is compact.

(3') COROLLARY. *If* X *is locally compact and* $f:X \Rightarrow Y$ *is a monotone mapping, for any compact set* K *in* X, $f^{-1}f(K)$ *is compact. If* K *is a continuum so also is* $f^{-1}f(K)$.

(3'') COROLLARY: *A monotone mapping on a locally compact space is compact if and only if each compact set in* Y *has a compact trace.*

It may be noted that the monotoneity condition in this theorem is quite essential, even though point inverses are compact. For if X is the part of the graph of the parabola $y^2 = x$ for which $-1 < y \leqq 1$, the vertical projection of X onto the interval $I = [0, 1]$ meets all conditions except monotoneity and yet I has a compact trace but a noncompact inverse.

4. An application. Invariance of the plane. One of the most outstanding results in the field of topology asserts that if we decompose the Euclidean plane into an upper semi-continuous collection of disjoint continua, not separating the plane, the decomposition space is itself a topological plane and thus homeomorphic with the original plane. This theorem (R. L. Moore [13]) still has no satisfactory simple extension to higher dimensional Euclidean spaces. We indicate next how a form of this fine theorem may be obtained from the sequence of results just discussed.

For the purposes of this discussion we define an S^2 (2-sphere) as a nondegenerate unicoherent Peano continuum containing a simple closed curve and separated irreducibly by every such curve it contains. Similarly an $\mathcal{E}^2$ is any locally compact Hausdorff space whose 1-point compactification is an S^2 or, equivalently, which is homeomorphic with the complement of some point on some S^2. That any S^2 or $\mathcal{E}^2$ is homeomorphic with an ordinary 2-sphere or plane respectively will not be used or proved in this discussion, although this is the case as is well known. Indeed the condition of unicoherence is redundant and imposed only for convenience in this discussion.

We first sketch briefly a proof for

(4) THEOREM. *The property of being an* S^2 *is invariant under nonconstant, nonseparating monotone mappings onto Hausdorff spaces.*

It is necessary first to establish the following sequence (i)–(v) of results in any S^2.

(i) JORDAN CURVE THEOREM: Every simple closed curve J on S^2 separates S^2 into exactly two components and is the boundary of each of these. The only thing left to prove here is that there are just two components of $S^2 - J$. This is accomplished by constructing an auxiliary simple closed curve J' using two disjoint arcs of J and arcs in two of the components of $S^2 - J$. Existence of a third component of $S^2 - J$ would force all of J into the closure of a single component of $S^2 - J'$ so that J' could not separate S^2.

(ii) θ-CURVE THEOREM: For any θ-curve θ on S^2, $S^2 - \theta$ has exactly three components and each of these is bounded by a unique pair of the edges of θ. This is an easy consequence of the Jordan Curve Theorem.

(iii) BOUNDARY CURVE THEOREM (Torhorst): The boundary B of any component R of the complement of a Peano continuum M on S^2 is itself a Peano continuum. This is proven by showing that B is a continuum by unicoherence of S^2 and, if it is not locally connected, a θ-curve θ can be constructed in M so that each of its 3 open edges can be connected to B without meeting either of the other two edges—thus contradicting the fact that R lies in just one of the three complementary regions of θ on S^2.

(iv) SEPARATION THEOREM: If A and B are continua on S^2 so that $A \cdot B$ is totally disconnected and both $A_0 = A - A \cdot B$ and $B_0 = B - A \cdot B$ are connected, then S^2 contains a simple closed curve J with $J(A+B) = A \cdot B$ which separates A_0 and B_0 in S^2. Further, if B does not separate S^2, then, for any $\sigma > 0$, J can be chosen so that the component of $S^2 - J$ containing B_0 lies in the σ-neighborhood of B.

This is proven by covering B_0 with the union of a sequence of regions whose closures are Peano continua not meeting A and so that only finitely many are not in the ϵ-neighborhood of $A \cdot B$ for each $\epsilon > 0$. Then by unicoherence of S^2, $\overline{Q}$ is a Peano continuum. It then follows that the boundary J of the component R of $S^2 - \overline{Q}$ containing A_0 meets all our requirements. (Note that in case B doesn't separate S^2, for a given σ we may suppose A_0 contains all points of S^2 not in the σ-neighborhood of B.)

To see this we first note that J must be either an arc or a simple closed curve. For if J contains a simple closed curve C, a simple θ-curve construction along with (ii) shows that J must reduce to C. On the other hand if J were a dendrite it could have no branch point because again a θ-curve construction in $J+R$ would likewise lead to a contradiction of (ii).

The proof is then completed by applying the part of the conclusion already established to show that J cannot be a simple arc because its end points would be accessible from R so that J would lie on a simple closed curve in S^2 and hence could not separate S^2. That any point p of J would be accessible from R follows from the fact that $R+p$ is locally connected because, by the part already proven, p lies in an arbitrarily small region in S^2 bounded by an arc or a simple closed curve C which meets J in at most two points so that $R \cdot C$ would have at most three components.

(v) DYGON THEOREM. A dygon is a continuum D which is the union of two continua A and B meeting in exactly two points x and y (the vertices of D) and so that $A_0 = A - (x+y)$ and $B_0 = B - (x+y)$ (the edges of D) are connected sets.

The Dygon Theorem asserts that: *Every dygon D on S^2 separates S^2; and if the boundary of any component of $S^2 - D$ meets both edges of D it also contains both vertices of D.*

The proof is an easy application of (iv) and (ii). For (iv) yields a simple closed curve J in S^2 separating the edges of D and meeting D in just its vertices x and y. The two open arcs of J from x to y are then separated in S^2 by D, since otherwise one can easily construct a θ-curve containing J and meeting D only in $x+y$. This contradicts (ii).

The Dygon Theorem yields the following key lemma:

LEMMA. *Let $\phi: S \Rightarrow \Sigma$ be monotone, where S is an S^2. Any simple closed curve J in Σ containing two distinct points x and y with unique inverses separates Σ; and if the boundary of any component of $\Sigma - J$ meets both open arcs xy of J, it must also contain both x and y.*

Proof of Lemma. Since ϕ is monotone, $\phi^{-1}(J)$ is a dygon D with vertices $x' = \phi^{-1}(x)$ and $y' = \phi^{-1}(y)$. Hence D separates S so that J separates Σ by monotoneity of ϕ. Likewise, if Q is any component of $\Sigma - J$ whose boundary B meets both open arcs xy of J, the boundary of $\phi^{-1}(Q)$ meets both edges of D and thus contains both x' and y', so that B necessarily contains both x and y.

With this lemma we are now in position to prove (4), the Sphere Invariance Theorem:

The property of being an S^2 is invariant under nonconstant monotone nonseparating mappings onto a Hausdorff space.

Proof. Let $f: S \to \Sigma$ be such a mapping where S is an S^2. Then Σ is a locally connected unicoherent continuum since these properties are invariant under monotone mappings. Also Σ has no cut point since f is nonseparating, and thus Σ contains a simple closed curve.

Thus if Σ is not an S^2 it must contain a simple closed curve J such that $\Sigma - J$ either (1) is connected or (2) contains a component whose boundary is an arc ab of J. Let x and y be distinct points of J, chosen so that they separate a and b on J in case (2) holds.

Decompose S into the continua $f^{-1}(x)$, $f^{-1}(y)$ and individual points of $S - f^{-1}(x) - f^{-1}(y)$. Let $\phi: S \Rightarrow S'$ be the natural mapping of this decomposition. Then ϕ is monotone, nonconstant and nonseparating, so that S' is a unicoherent locally connected continuum containing a nondegenerate simple closed curve. Also, if J' is any simple closed curve in S', and a_1 and b_1 are any two points of J', there always exist points x_1 and y_1 with unique ϕ inverses so that x_1 and y_1 separate a_1 and b_1 on J'. Hence, by the lemma, J' separates S', and no simple arc ab of J' can separate S'. Accordingly S' is an S^2.

Now define $g: S' \Rightarrow \Sigma$ by $g(p) = f\phi^{-1}(p)$ for $p \in S'$. It is verified at once that g

is monotone and nonseparating. Thus since $g^{-1}(x)$ and $g^{-1}(y)$ are single points of S', the lemma applied to g shows (a) that J separates Σ so that case (1) cannot arise and (b) that there can be no component of $\Sigma - J$ whose boundary is an arc ab of J as required by case (2). Thus the supposition that Σ is not an S^2 leads to a contradiction.

As an easy consequence we now get

(5) THEOREM. *The property of being an $\mathcal{E}^2$ is invariant under quasi-compact monotone nonseparating mappings onto Hausdorff spaces.*

For let $\phi: X \Rightarrow Y$ be such a mapping where X is an $\mathcal{E}^2$. Then ϕ is closed and Y is locally compact by (1). Thus by (2), ϕ has a continuous extension $\phi^*: X^* \Rightarrow Y^*$ to the 1-point compactification X^* of X onto that of Y. Then X^* is an S^2 and ϕ^* is nonconstant, nonseparating and monotone and Y^* is a Hausdorff space. Thus by (4), Y^* is an S^2; and hence Y itself is an $\mathcal{E}^2$.

REMARK. In view of the well-known fact, not proven here (see above), that all S^2's are homeomorphic with each other, as are also all $\mathcal{E}^2$'s, as a consequence of (4) and (5) we have the classical result of R. L. Moore which asserts that:

The decomposition space for any nontrivial upper semicontinuous decomposition of the plane, or 2-sphere X into disjoint continua not separating X, is homeomorphic with X.

5. The compact part of a mapping. For any mapping $f: X \Rightarrow Y$ we define (i) Q as the union of the interiors of sets in Y having a compact inverse and $P = f^{-1}(Q)$, (ii) Q' as the union of the interiors of the images of all compact sets in X. It results at once that P, Q and Q' are open sets, and that the mapping $f|P: P \Rightarrow Q$ is compact. Also, using Theorem (3) above, we get at once

THEOREM A. *Let X be locally compact. If f is monotone then $Q = Q'$. If f is (1-1), Q is exactly the set of points of continuity of f^{-1}.*

THEOREM B. *Let X be locally compact and have a countable base and suppose Y is a complete metric space. Then Q' is dense in Y. Further, if f is monotone $Q = Q'$; and if f is 1-1, f maps P topologically onto Q.*

6. One-to-one Mappings. These are always of special interest. One is naturally concerned with conditions under which they are homeomorphisms and with the extent to which they preserve the topological structure of their domain spaces even when they are not homeomorphisms.

We have already noted that $w = e^{i\theta}$, $0 \leq \theta < 2\pi$, maps the half open interval onto the circle $|z| = 1$ in 1-1 fashion and also that the plane can be mapped similarly onto the pinched 2-sphere obtained by identifying its poles. In both of these cases the range space is compact and the domain locally compact. Of course if the domain space is compact, the mapping is a homeomorphism. Also any quasi-compact 1-1 mapping is a homeomorphism.

In general the 1-1 continuous image of even a simple noncompact space may be quite different in structure from that of the original. For example such an image of the line may fail to be locally connected, as is well known. If we consider only 1-1 mappings onto locally connected spaces, significant results have been found by McAuley and Lelek [14] and Jones [15] in case the domain is a line or a plane. The only possible 1-1 locally connected continuous images of the line, for example, are the line, figure eight, noose, dumbbell or θ-curve.

In this particular case it may be noted that if we impose the additional condition of unicoherence on the image space, then the only possible image of the line is the topological line itself. Further, in this case, the mapping itself must be topological. This and other related observations and results led to the conjecture at one time that a 1-1 mapping from a locally connected locally compact connected separable metric space onto a unicoherent space of the same sort would necessarily be a homeomorphism. Indeed such a result was asserted [16] and thought to be proven at one time. However, examples were given by Kenneth Whyburn [17] and L. C. Glaser [18] showing that this is not true and, indeed, not even when the image space is E^n for $n \geqq 3$. The case of 1-1 mappings of such spaces onto E^2 has been studied by E. Duda [19] who has shown that these mappings are homeomorphisms provided the domain spaces satisfy some simple additional restrictions.

A fairly simple example in which the image is a unicoherent Peano continuum (but not a Euclidean space) may be constructed as follows. Let $\triangle$ be an equilateral triangle ABC including its interior and let O be the center of $\triangle$. Then add to $\triangle$ three solid triangles $A'OB$, $B'OC$, and $C'OA$ meeting $\triangle$ in just the segments OB, OC, OA respectively, and meeting each other by pairs only in O. Let X' be the union of these 4 solid triangles, and let X' be mapped onto a space Y by identifying A' with A and mapping $A'B$ linearly onto AB; similarly $B'C$ is mapped onto BC and $C'A$ onto CA. Otherwise the mapping is 1-1. Thus Y consists of $\triangle$ plus three open pockets created by identifying points on $A'B$ with corresponding points on AB, and similarly for $B'C$ and $C'A$. Finally, we delete from the domain X' of the mapping all points on the periphery of the original triangle $\triangle = ABC$. Let X be the resulting space $X = X' - $ (periphery of $\triangle$), and $f: X \Rightarrow Y$ the restriction of the mapping to X. Then f is 1-1 and continuous, X is locally compact, connected and locally connected, and Y is a unicoherent Peano continuum, but of course f is not a homeomorphism. (Indeed, Y is compact but X is not.)

It is not enough, in general, to assume X and Y homeomorphic with each other in order to assure that every 1-1 mapping of X onto Y will be a homeomorphism, even when these spaces are locally connected generalized continua (= locally compact, connected separable metric spaces). It is true, however, that in case both X and Y are Euclidean spaces, any 1-1 mapping of X onto Y must be a homeomorphism. This conclusion is obtainable as a consequence of the Brouwer Theorem on invariance of openness in Euclidean spaces. Indeed we can get a somewhat better result by formulating the Brouwer Property in a broader setting as follows.

A space X is said to have the Brouwer Property [12], provided any subset of X which is homeomorphic with some open set in X is itself open in X. All Euclidean spaces have this property by Brouwer's Theorem, as do also all Euclidean manifolds. However, a large class of spaces which are not Euclidean also have this property. For example, the set obtained by taking the 1-point compactification of a surface of infinite genus. Also the generalized closed manifolds of Wilder have the Brouwer Property although they are not all Euclidean by any means. A straightforward argument suffices to prove

THEOREM. *If X and Y are homeomorphic locally compact spaces having the Brouwer Property, any 1-1 mapping of X onto Y is a homeomorphism.*

Now in case X and Y are Euclidean spaces one of which is mappable onto the other by a 1-1 mapping, it results from Theorem B in section 5 above that X and Y are of the same dimension and thus are homeomorphic with each other. Thus any 1-1 mapping of one Euclidean space onto another is a homeomorphism.

For further examples and results in connection with the Brouwer Property the reader is referred to a study made by E. Duda [20].

7. The compactness problem. Results of this sort in the 1-1 case led the author many years ago to formulate and study the problem of determining conditions under which a mapping of a Euclidean space E^n onto itself is necessarily compact. The cases $n = 1, 2$, were solved and appeared in a paper published in 1959 [12]. For $n = 1$ we need only assume compactness of point inverses and for $n = 2$ it is enough to assume them to be continua, i.e., assume the mapping monotone. Higher dimensional cases of the problem present considerably greater difficulty. We now consider these various cases in some detail.

(i) $n = 1$. *If X and Y are lines (E^1), a mapping $f: X \Rightarrow Y$ is compact if (and only if) it has compact point inverses.*

Let f have compact point inverses and take any compact set K in Y. Let ab be an interval in X whose interior contains $f^{-1}(\alpha + \beta)$, where $\alpha\beta$ is an interval in Y whose interior contains K. Then neither $-\infty a$ nor $b\infty$ can meet $f^{-1}(K)$. For if, say, $-\infty a$ meets $f^{-1}(K)$, $f(-\infty a)$ lies wholly in $\alpha\beta$ so that $f(-\infty a + ab)$ is bounded. Thus $f(b\infty)$ would meet both $-\infty\alpha$ and $\beta\infty$ and thus would contain α and β which is impossible.

In terms of real valued functions this result says that if f is a real function of a real variable continuous everywhere on E^1 which takes each real value at least once, but only on a compact set, then

$$\lim_{x \to \infty} f(x) = \pm \infty, \ \lim_{x \to -\infty} f(x) = \mp \infty.$$

(ii) $n = 2$. In the case of a mapping of a plane onto a plane it is not enough to have compact point inverses in order to make the mapping compact. For the function

$$w = \left(\frac{x}{x+1} + iy\right)^2,$$

i.e., $w = z'^2$, where

$$z' = \frac{x}{x+1} + iy,$$

maps the complex plane onto itself and yet f takes on each value either one or two times. However, as indicated earlier, it is sufficient (though not necessary, of course) that point inverses be continua.

THEOREM. *If X and Y are planes, any monotone mapping $f: X \Rightarrow Y$ is compact (and nonseparating).*

Proof. The map f generates a decomposition of X into continua. Let Y' denote the decomposition space and $\phi: X \Rightarrow Y'$ the decomposition map. There is the one-to-one map $h: Y' \Rightarrow Y$ with $h\phi = f$. Now ϕ is quasi-compact, hence compact by (1) of section 3. We therefore have a unique monotone map $\phi^*: X^* \Rightarrow Y'^*$ as in (2) of section 3. By the general form of Moore's Theorem, Y'^* is a cactoid.

Case 1: Y'^* is a sphere. In this case Y' is a plane. Hence, the one-to-one map $h: Y' \Rightarrow Y$ must be open and a homeomorphism by Brouwer's Theorem. Hence f is compact in this case.

Case 2: Y'^* contains a sphere as a proper subset. No sphere can map one-to-one into the plane, hence Y' must contain a plane E which has limit points of $Y' - E$. Then h maps E homeomorphically onto an open subset of Y. Since E contains limit points of $Y - E$, it is impossible to extend $h: E \rightarrow Y$ to a continuous one-to-one map of Y' into Y.

Case 3. Y'^* is a dendrite. By Theorem B, h must be a homeomorphism of some nonempty open subset onto an open subset of Y. But every open subset of a dendrite has cut points, whereas no open subset of the plane has cut points.

Hence f is compact. The mapping $f^*: X^* \Rightarrow Y^*$ is clearly nonseparating, hence so also is $f: X \Rightarrow Y$.

Since the mapping f in this theorem turns out to be automatically nonseparating, in view of results on invariance of the plane discussed earlier we have the

COROLLARY. *Let $f: X \Rightarrow Y$ be a monotone mapping where X is a plane and Y is a Hausdorff space. Then Y is a topological plane if and only if f is compact and nonseparating.*

(iii) $n \geq 3$. These cases were discussed by the author at the Georgia Conference on Topology of 3-manifolds in 1961 (see Proceedings of this Conference, by M. K. Fort, Prentice-Hall, 1961). It was conjectured that a 1-monotone mapping of E^3 onto E^3 is always compact and, possibly, an $(n-2)$-monotone mapping of E^n onto E^n is compact—a mapping being *r-monotone* provided its point inverses have trivial homology groups in dimension $\leq r$. The conjecture in this form would reduce to exactly the above result in (ii) for $n = 2$ and to the conjectured one for $n = 3$. This problem and conjecture attracted considerable attention and effort on the part of those in attendance at the conference and others who learned of it later. It was shown by Connell at that time that acyclic maps of E^n onto E^n are always compact, where "acyclic" means that homology groups of *all* dimensions of point inverses are trivial. See the above mentioned Proceedings.

In 1965 a positive solution to the problem, verifying the conjecture for general n in the form given above, was published by Väisälä, along with a number of related results. Methods of proof required, however, are much more algebraic and less elementary than needed in the cases $n = 1$, 2. An independent proof of this same general result: *$(n-2)$-monotone mappings of E^n onto E^n are compact* was found by Conner and Jones and distributed in manuscript form. Their methods differ from those of Väisälä in that they depend more directly on homology theory and exact sequences rather than on the topological index. However, both proofs are far from elementary; and a simple direct set-theoretic type of proof for this fine theorem for general n would be highly desirable and would represent a valuable contribution to mathematical knowledge.

The question as to the existence of noncompact monotone ($= 0$-monotone) mappings of E^n onto E^n for $n \geq 3$ has been studied in detail by Glaser [18]. He has shown that such maps do indeed exist for $n \geq 4$ and has further results on existence of monotone maps of E^k onto E^n, where k and n are not necessarily equal. At the 1969 Georgia Conference, R. H. Bing gave an example of a monotone mapping of E^3 onto E^3 which is not compact.

Also it would be of interest to determine just what properties of Euclidean spaces are essential in making these results on compactness valid. Can one formulate sufficient conditions on a space, perhaps in terms of its homology or homotopy groups or contractibility or retractability into subsets which will suffice to yield similar results on compactness to the ones discussed here for Euclidean spaces? Possibly the Brouwer Property could be useful in this connection.

8. Compactification of mappings. The existence of various forms of compactifications of a topological space X (i.e., the topological imbedding of X in a larger compact space in which X is dense) such as the one-point compactification, the Stone-Čech compactification, etc., leads naturally to the question as to whether certain types of mappings can be compactified in some similar sense. This question arose naturally also in connection with a result of Vainstein [5] to the effect that *any closed mapping has a partial mapping* (restriction) *which is compact and in which the image space is the same as the image of the original domain space.* The usual dual relationship between open and closed (sets, mappings, etc.) led the author to anticipate that open mappings may be related in some analogous way to compact mappings. More particularly, it was anticipated that if the domain space were suitably augmented, the mapping could be extended so as to become compact, so that the given mapping would be exhibited as a partial mapping of a compact one.

This was found to be fully correct. In a paper published in 1953 [21] it was shown that in a suitably constructed space unifying both the domain and range spaces, a compact mapping can be defined which is topologically equivalent to the given one on the prototype of the original domain space. This turned out to be true for arbitrary mappings, not just for open ones. Thus it was shown that any mapping from one Hausdorff space to another is topologically equivalent to a partial mapping of a compact one (actually a retraction) in the unified space. The extended mapping is open when the original one was open; and the unified space is separable and metrizable when the given spaces are locally compact, separable and metric. More recently H. Bauer [22] had occasion to use such a result in his studies on measure preservation of mappings and, independently, he developed and applied almost exactly the same extension to show that any mapping between locally compact spaces is a partial mapping of a conservative one.

Sketched briefly, we begin with an arbitrary mapping $f: X \to Y$, where X and Y are disjoint Hausdorff spaces, and define the *unified space* Z to consist of all points in the union of X and Y with a topology in which a set $Q \subset Z$ is open if and only if

(i) *$Q \cdot X$ and $Q \cdot Y$ are open in X and Y respectively, and*
(ii) *for any compact set $K \subset Q \cdot Y$, $f^{-1}(K) \cdot (X - Q)$ is compact.*

It turns out that the injection mappings of X and Y into Z are open and closed respectively and thus are homeomorphisms. Further, if we define $r(x) = f(x) \in Z$ for $x \in X$ and $r(x) = x \in Z$ for $x \in Y$, then r is continuous and compact and hence is a compact retraction of Z onto Y. Since $r | X$ is f followed by the injection of Y into Z it is therefore topologically equivalent to f. If f is an open mapping, so also is r. If X and Y are locally compact, Z is locally compact and Hausdorff; and, in this case, if X and Y are also separable and metric so also is Z.

In general, the set X is not necessarily dense in the space Z, i.e., $\tilde{X} \neq Z$ unless f is everywhere noncompact, where $\tilde{X}$ denotes the closure of X in Z. However, X is dense in $\tilde{X}$ of course and the partial mapping

$$r | \tilde{X} : \tilde{X} \to Y$$

is a compact extension of f to $\tilde{X}$. Note also that since r is a retraction and $\tilde{X} - X \subset Y$, $r | (\tilde{X} - X)$ is the identity mapping. The original mapping, considered as a mapping of X into its image, i.e., $f: X \to f(X)$, is a compact mapping if and only if both r and r^{-1} reduce to the identity on the set $\tilde{X} - X$. The important special case of a 1-1 mapping $f: X \to Y$ which is compact relative to $f(X)$, i.e., in case f is a homeomorphism of X onto $f(X)$, is of special interest here. For if f is a homeomorphism of X onto $f(X)$, it turns out that r extends f to a homeomorphsim of $\tilde{X}$ onto $\overline{f(X)}$.

The unified space for a mapping has been further studied by R. Dickman [23] who developed, in particular, conditions for unicoherence of this space and implications of its unicoherence so far as the mapping is concerned.

The process of compactifying a mapping has been studied from a more general viewpoint by Dickman [24], Cain [25], and others, and quite interesting results obtained. A compactification of a mapping $f: X \Rightarrow Y$ is defined as a pair (X^*, f^*) where X^* is a Hausdorff space containing X as a dense subspace and $f^*: X^* \Rightarrow Y$ is a closed continuous extension of f from X^* onto Y having compact point inverses. Thus f^* is a compact mapping and $f^* | X = f$.

Dickman [26] has developed the concept of maximal and minimal compactifications of spaces and, correspondingly, of mappings. Cain [27] has developed two different procedures for characterizing all possible compactifications for a given mapping. In each case there is associated with each possible compactification X' of the domain space X a compactification of the mapping $f: X \Rightarrow Y$ in a unique way so that the mapping compactification is determined by the space compactification and so that every possible mapping compactification is obtainable by the given process from a space compactification.

The first of these procedures employs filter space compactifications of spaces as developed by Wagner. The second uses in a similar way the ring compactifications of spaces provided by the structure space of the subring R of the ring of all bounded continuous real valued functions on the domain space X. This structure space has as points the maximal ideals in R and is a Hausdorff compactification of X. (See Wagner [28].)

References

1. B. L. McAllister, Cyclic elements in topology, A history, this MONTHLY, 73 (1966) 337–350.
2. R. H. Bing (editor), Summer Institute on Set Theoretic Topology, Madison, 1955.
3. M. K. Fort, Jr. (editor), Topology of 3-manifolds and related topics, Proc. University of Georgia Conference, Prentice-Hall, 1962.
4. R. L. Wilder, Topology of manifolds, A.M.S. Colloq. Publ., 32 (1949).
5 I. A. Vainstein, On closed mappings, Moskov. Gos. Univ. Uc. Zap. 155, Mat., 5 (1952) 3–53.
6. G. T. Whyburn, Open mappings on locally compact spaces, Mem. Amer. Math. Soc., 1 (1950) 24.
7. ———, Directed families of sets and closedness, Proc. N.A.S. no. 3, 54 (1965) 688–692.
8. ———, Open and closed mappings, Duke Math. J., 17 (1950) 69–74.
9. P. Alexandroff and H. Hopf, Topologie, vol. 1, Springer-Verlag, 1935.
10. G. T. Whyburn, Mappings on inverse sets, Duke Math. J., 23 (1956) 237–240.
11. ———, Notes on general topology and mapping theory, mimeographed notes, U. of Virginia, 1965.
12. ———, Compactness of certain mappings, Amer. J. Math., 81 (1959) 306–314.
13. R. L. Moore, Concerning upper semi-continuous collections of continua, Trans. Amer. Math. Soc., 27 (1925) 416–428.
14. A. Lelek and L. F. McAuley, On hereditarily locally connected spaces and one-to-one continuous images of a line, Colloq. Math., 17 (1967) 319–324.
15. F. Burton Jones, On a plane one-to-one map of a line, Colloq. Math., 19 (1968) 231–233.
16. V. V. Proizvolov, One-to-one mappings onto metric spaces, Dokl. Akad. Nauk SSSR, 158 (1964) 788–789 and 1286–1287.
17. Kenneth Whyburn, A non-topological 1-1 mapping onto E^3, Bull. Amer. Math. Soc., 71 (1965) 533–537.
18. L. C. Glaser, Dimension lowering monotone non-compact mappings of E^n, Fund. Math., 58 (1966) 177–181.
19. Edwin Duda, A theorem on one-to-one mappings, Pac. J. Math., 19 (1966) 253–257.
20. ———, Brouwer property spaces, Duke Math. J., 30 (1963) 647–660.
21. G. T. Whyburn, A unified space for mappings, Trans. Amer. Math. Soc., 74 (1953) 344–350.

22. H. Bauer, Konservative Abbildungen lokalkompacter Räume, Math. Ann., **138** (1959) 398–427.

23. R. F. Dickman, Unified spaces and singular sets of mappings of locally compact spaces, Fund Math., **62** (1968) 103–123.

24. ———, On closed extensions of functions, Proc. N.A.S., **62** (1969) 326–332.

25. G. L. Cain, Jr., Compact and related mappings, Duke Math. J., **33** (1966) 639–645.

26. R. F. Dickman, Compactifications and real compactifications of arbitrary topological spaces, to appear.

27. G. L. Cain, Jr., Compactifications of mappings, Notices, Amer. Math. Soc., **16** (1969) 277.

28. F. J. Wagner, Notes on compactification I, Nederl. Akad. Wetensch. Proc. Ser. A, **60** (1957) 171–176.

Bibliography

Part A

§II

P. S. Alexandroff and Heinz Hopf, *Topologie I*, Berlin: Springer-Verlag, 1935.

Maurice Fréchet, *Les Espaces Abstraits, Paris*: Monographies Borel, 1928.

Felix Hausdorff, *Gründzuge der Mengenlehre*, Leipzig: Veit, 1914.

Leopold Vietoris, Stetige Mengen, *Monatshefte für Matematik und Physik*, vol. **3** (1921), p. 173.

Heinrich Tietze, Beitrage zur allgemeinen Topologie I, *Mathematische Annalen*, vol. **88** (1923).

Ernst Lindelöf, Sur quelques points de la theorie des ensembles, *Comptes Rendus Hebdomadaires des Seances de L'Académie des Sciences, Paris*, vol. **137** (1903).

A. N. Tychonoff, Über einen Metrizationsatz von P. Urysohn, *Mathematische Annalen*, vol. **95** (1926).

§IV

Bernard Bolzano, *Abhandlungen der Konigliche böhmuschen Gesellschaft der Wissenschaften*, 1817.

Emil Borel, Sur quelques points de la théorie des functions, *Annales scientifiques de L'Ecole Normale Supérieure*, vol. **(3)** 12 (1895), p. 51.

T. H. Hildebrandt, The Borel theorem and its generalizations, *Bulletin of the American Mathematical Society*, vol. **32** (1926), pp. 423–474.

§V

P. Urysohn, Über die Machtigkiet der zusammenhängenden Mengen, *Mathematische Annalen*, vol. **94** (1925), pp. 309–315.

§VI

Maurice Fréchet, Sur quelques points du calcul fonctionnel, *Rendiconti del Circolo Matematico di Palermo*, vol. **22** (1906).

P. Urysohn, Über die Metrization der kompakten topologischen Räume, *Mathematische Annalen*, vol. **92** (1924), pp. 275–293.

§VIII

P. Painlevé, *Comptes Rendus Paris*, vol. **148** (1909), p. 1156.

§X

N. J. Lennes, Curves in non-metrical analysis situs with applications in the calculus of variations, *American Journal of Mathematics*, vol. **33** (1911), pp. 287–326.

§XI

L. Zoretti, Sur les functions analytiques uniformes, *Journal de Mathématiques Pures et Appliquées*, (6), vol. **1** (1905), pp. 9–11.

R. G. Lubben, Concerning limiting sets in abstract spaces, *Transactions of the American Mathematical Society*, vol. **30** (1928), pp. 668–685.

§XII

S. Janiszewski, Sur les continus irréductibles entre deux points, *Journal de l'Ecole Polytechnique* (2), vol. **16** (1912), pp. 76–170.

A. M. Mullikin, Certain theorems relating to plane connected point sets, *Transactions of the American Mathematical Society*, vol. **24** (1922), pp. 144–162.

R. L. Moore, Concerning the sum of a countable number of mutually exclusive continua in the plane, *Fundamenta Mathematica*, vol. **6** (1924), pp. 189–202.

W. Sierpinski, Un théoréme sur les ensembles fermes, *Bulletin de l'Académie des Sciences, Cracovie* (1918), pp. 49–51.

§XIII

L. E. J. Brouwer, Over de structur der perfect punktmerzamelingen, *Akademie Versammlungen*, vol. **18** (1910), pp. 833–842; vol. **19** (1911), pp. 1416–1426. See also *Proceedings, Akademie van Wetenschappen, Amsterdam*, vol. **14** (1911), p. 138.

S. Janiszerski, loc. cit.

§XIV

H. Hahn, Über die Kompomenten offenen Mengen, *Fundamenta Mathematicae*, vol. **2** (1921), pp. 189–192.

K. Kuratowski, Une définition topologique de la ligne de Jordan, *Fundamenta Mathematicae*, vol. **1** (1920), pp. 40–43.

R. L. Moore, Report on continuous curves from the viewpoint of analysis situs, *Bulletin of the American Mathematical Society*, vol. **29** (1923), pp. 289–302.

§XV

R. L. Moore, Concerning connectedness im kleinen and a related property, *Fundamenta Mathematicae*, vol. **3** (1922), pp. 232–237.

W. Sierpinski, Sur une condition pour qu'un continu soit une courbe jordanienne, *Fundamenta Mathematicae*, vol. **1** (1920), pp. 44–60.

§XVI

G. T. Whyburn, Open and closed mappings, *Duke Mathematics Journal*, vol. **17** (1950), pp. 69–74.

G. T. Whyburn, On quasi-compact mappings, *Duke Mathematics Journal*, vol. **19** (1952), pp. 445–446.

§XVII

Maurice Fréchet, loc. cit. §VI.

Felix Hausdorff, *Mengenlehre*, Leipzig: de Gruyter, 1927.

§XVIII

P. Alexandroff and P. Urysohn, Mémoire sur les espaces topologiques compacts, *Verhandelingen der Akademie van Wetenschappen, Amsterdam*, vol. **14** (1929), pp. 1–96.

H. Hahn, Mengentheoretische Characterisierung der stetige Kurven, *Sitzungsberichte, Akademie der Wissenschaften, Vienna*, vol. **123** (1914), p. 2433.

S. Mazurkiewicz, Sur les lignes de Jordan, *Fundamenta Mathematicae*, vol. **1** (1920), pp. 166–209.

§XIX

R. L. Moore, Concerning simple continuous curves, *Transactions of the American Mathematical Society*, vol. **21** (1920), pp. 333–347.

§XX

K. Kuratowski, *Topologie I*, Warsaw, 1933.

S. Mazurkiewicz, loc. cit.

K. Menger, Zur Begründung einer axiomatichen Theorie der Dimension, *Monatshefte für Mathematik und Physik*, vol. **36** (1929), pp. 193–218.

R. L. Moore, Foundations of Point Set Theory, *American Mathematical Society Colloquium Publications*, vol. **13**, 1932; revised 1962.

R. L. Moore, On the foundations of plane analysis situs, *Transactions of the American Mathematical Society*, vol. **17** (1916), pp. 131–164.

H. Tietze, Über stetige Kurven, Jordansche Kurvenbogen und geschlossene Jordan Kurven, *Mathematische Zeitschrift*, vol. **5** (1919), pp. 284–291.

Appendix I

G. T. Whyburn, A note on spaces having the S property, *American Journal of Mathematics*, vol. **54** (1932), pp. 536–538.

Appendix II

W. L. Ayres, Concerning continuous curves in metric space, *American Journal of Mathematics*, vol. **51** (1929), pp. 577–594.

W. L. Ayres, Continuous curves which are cyclically connected, *Bulletin International de l'Académie Polonaise des Sciences et des Lettres, Classe des Sciences Mathématiques et Naturelles*, A (1928), pp. 127–142.

G. T. Whyburn, Cyclicly connected continuous curves, *Proceedings of the National Academy of Science*, vol. **13** (1927), pp. 31–38.

G. T. Whyburn, Concerning the structure of a continuous curve, *American Journal of Mathematics*, vol. **50** (1928), pp. 167–194.

G. T. Whyburn, Cyclic elements of higher order, *American Journal of Mathematics*, vol. **56** (1934), pp. 133–146.

G. T. Whyburn, On the structure of continua, *Bulletin of the American Mathematical Society*, vol. **42** (1936), pp. 49–73.

G. T. Whyburn, *Analytic Topology*, American Mathematical Society Colloquium Publications, vol. **28**, Providence: American Mathematical Society, 1942.

Part B

§I

P. S. Alexandroff, Über die Metrization der in kleinen kompakten topologischen Räume, *Mathematische Annalen*, vol. **92** (1924), pp. 294–301.

E. Cech, On bicompact spaces, *Annals of Mathematics*, vol. **38** (1937), pp. 823–844.

M. H. Stone, Applications of the theory of Boolean rings to general topology, *Transactions of the American Mathematical Society*, vol. **41** (1937), pp. 375–481.

A. N. Tychonoff, Über die topologische Erweiterung von Räumen, *Mathematische Annalen*, vol. **102** (1929), pp. 544–561.

A. N. Tychonoff, Über einen Functionenraum, *Mathematische Annalen*, vol. **111** (1935), pp. 762–766.

§II

P. Alexandroff, Über stetige Abbildungen kompakter Räume, *Mathematische Annalen*, vol. **92** (1927), pp. 555–571.

B. v. Kerékjártó, Involutions et surfaces continues, *Szeged, Acta Litterarum ac Scientiarum*, vol. **3** (1927), pp. 49–67.

K. Kuratowski, Sur les décompositions semi-continues d'espaces métriques compacts, *Fundamenta Mathematicae*, vol. **11** (1928), pp. 169–185.

R. L. Moore, Concerning upper semi-continuous collections of continua, *Transactions of the American Mathematical Society*, vol. **27** (1925), pp. 416–428.

R. L. Moore, loc. cit., Part A, §XX

L. Vietoris, Über stetige Abbildungen einen Kugelfläche, *Proceedings, Akademie van Wetenschappen, Amsterdam*, vol. **29** (1926), pp. 443–453.

§III

H. Bauer, Verallgeminerung eines Faktorisierungsatzes von G. T. Whyburn, *Archiv der Mathematik*, vol. **10** (1959), pp. 373–378.

K. Stein, Analytical Zerlegungen-Komplexen Räume, *Mathematische Annalen*, vol. **132** (1956), pp. 63–93.

G. T. Whyburn, loc. cit., Part A, §XVI.

§IV

K. Borsuk, Sur les prolongements des transformations continues, *Fundamenta Mathematicae*, vol. **28** (1937), pp. 99–110.

C. H. Dowker, Mapping theorems for non-compact spaces, *American Journal of Mathematics*, vol. **69** (1947), p. 232.

S. Eilenberg, Sur les transformations d'espaces métriques en circonférence, *Fundamenta Mathematicae*, vol. **24** (1935), pp. 160–176.

H. Tietze, Über Funktionen, die auf einer abgeschlossenen Menge stetig sind, *Journal für Mathematik*, vol. **145** (1915), pp. 9–14.

§V

L. E. J. Brouwer, Beweis des Jordanschen Kurvensatz, *Mathematische Annalen*, vol. **69** (1910), pp. 169–175.

S. Eilenberg, loc. cit., Part B, §IV.

K. Kuratowski, Sur le continua de Jordan et le théoréme de M Brouwer, *Fundamenta Mathematicae*, vol. **13** (1929), pp. 307–318.

§VI

C. Jordan, *Cours d' Analyse*, 2nd edition, Paris, 1893, p. 92.

R. L. Moore, loc. cit., Part A, §XX.

R. L. Moore, loc. cit., Part B, §II.

R. L. Moore, Concerning upper semi-continuous collections, *Monatshefte für Mathematik und Physik*, vol. **36** (1929), pp. 81–88.

M. Torhorst, Über den Rand der einfach zusammenhängenden ebene Gebiete, *Mathematische Zeitschrift*, vol. **9** (1921), pp. 44–65.

L. Zoretti, loc. cit., Part A, §XI.

Index

Arc, simple 70
Arcwise connected 74

Basis for a topology 7
Bolzano−Weierstrass (B−W) set 14
Boundary (frontier) 41
Borel Theorem 14, 16
Brouwer reduction theorem 43, 44

Cartesian product 85
Cartesian product of two metric spaces 23
Cauchy−Schwartz Inequality 25
Cauchy sequence 55
Closed curve, simple 70
Closed function 51
Closed set 6
Closure of a set 6
Compact sets 14
Compactification 86
Complete enclosure of a space 55
Complete space 55
Completely regular 85
Component of a set 34
Conditionally compact 14
Connected set 34
Continuum 41
Convergent sequence 7
Convergence of a sequence of sets 31
Cut point 79
Cut point, local 79

Cyclic element 79
Cyclically connected 79

Decomposition, component 100
Decomposition, upper-semicontinuous 93
Decomposition, lower-semicontinuous 93
Decomposition spaces 93
Dendrite 111
Directed family 12
Directed family, convergence of 12
Directed family, cluster point of 12
Directed family, underfamily of 12
Distance function 22
Dygon 119
Dygon Theorem 127

ϵ-chain 34
End points of a simple arc 70
End point of a space 79
Equivalent metrics 112
Euclidean plane, E^n 23
Exponentially representable 107

Function 17
Function, onto 17
Function, one−to−one (1-1) 17
Function, continuous (=mapping) 17
Function, factorization of 100
Function space Y^X 55

Generalized continuum   46

Hausdorff Axiom   6
Hausdorff Maximality Principal   89
Hilbert parallelotrope   23
Hilbert space l_2, H   23
Homeomorphism   51
Homotopy Extension Theorem   109
Homotopic mappings   105

Irreducible continuum   43

Jordan Curve Theorem   120, 122

Light function   100
Limit point   6
Limit inferior   31
Limit superior   31
Limit theorem   31, 32
Lindelöf Theorem   8, 11
Locally compact   45
Locally connected   45

Metric space   22
Metrization Theorem   23, 24
Minkowski Inequality   25
Monotone function   93
Multicoherent   112

Nonseparating map   120
Normality axiom   7

Open function   51

Partially ordered set   89
Perfectly separable (2nd countable)   7
Plane Separation Theorem   120, 126
Projection function   85
Property S   48

Quasicompact function   51
Quotient space   93

Region   45
Regularity axiom   6
Relative topology   33

Semipolygon   119
Separable space   22
Separation of a set   34
Set   3
Sets, union of   3
Sets, intersection of   3
Sets, complement of   3
Sets, null   4
Sets, countable   4
Simple arc   70
Simple closed curve   70
Simply ordered chain   89
Stereographic projection   107

τ_1—axiom   6
θ−curve   111
Tietze Extension Theorem   108
Topological space   6
Topologically equivalent functions   100
Torhorst Theorem   124
Totally bounded space   55
Triangle inequality   22
Tychonoff Lemma   8, 11
Tychonoff space   85
Tychonoff Theorem   86, 90

Unicoherent space   111
Uniformly locally connected   48
Universal exponential representation
    property   111
Urysohn's Lemma   18, 20

Weakly monotone mapping   111
Well chained   34

Undergraduate Texts in Mathematics

Apostol: Introduction to Analytic
Number Theory.
1976. xii, 370 pages. 24 illus.

Childs: A Concrete Introduction to
Higher Algebra.
1979. xiv, 338 pages. 8 illus.

Chung: Elementary Probability Theory
with Stochastic Processes. Third Edition
1979. 336 pages.

Croom: Basic Concepts of Algebraic
Topology.
1978. x, 177 pages. 46 illus.

Fleming: Functions of Several Variables.
Second edition.
1977. xi, 411 pages. 96 illus.

Halmos: Finite-Dimensional Vector
Spaces. Second edition.
1974. viii, 200 pages.

Halmos: Naive Set Theory.
1974. vii, 104 pages.

Kemeny/Snell: Finite Markov Chains.
1976. ix, 210 pages.

Lax/Burstein/Lax: Calculus with
Applications and Computing,
Volume 1.
1976. xi, 513 pages. 170 illus.

LeCuyer: College Mathematics with
A Programming Language.
1978. xii, 420 pages. 126 illus. 64 diagrams.

Malitz: Introduction to Mathematical
Logic.
Set Theory - Computable Functions -
Model Theory.
1979. 255 pages. 2 illus.

Prenowitz/Jantosciak: The Theory of
Join Spaces.
A Contemporary Approach to Convex
Sets and Linear Geometry.
1979. Approx. 350 pages. 404
illus.

Priestley: Calculus: An Historical
Approach.
1979. 400 pages. 300 illus.

Protter/Morrey: A First Course in Real
Analysis.
1977. xii, 507 pages. 135 illus.

Sigler: Algebra.
1976. xi, 419 pages. 32 illus.

Singer/Thorpe: Lecture Notes on
Elementary Topology and Geometry.
1976. viii, 232 pages. 109 illus.

Smith: Linear Algebra
1978. vii, 280 pages. 21 illus.

Thorpe: Elementary Topics in
Differential Geometry.
1979. 256 pages. Approx. 115 illus.

Whyburn/Duda: Dynamic Topology.
1979. Approx. 175 pages. Approx. 20
illus.

Wilson: Much Ado About Calculus.
A Modern Treatment with Applications
Prepared for Use with the Computer.
1979. Approx. 500 pages. Approx. 145
illus.

Algebraic Topology: An Introduction
by **W. S. Massey**
(Graduate Texts in Mathematics, Vol. 56)
1977. xxi, 261p. 61 illus. cloth

Here is a lucid examination of algebraic topology, designed to introduce advanced undergraduate or beginning graduate students to the subject as painlessly as possible. *Algebraic Topology: An Introduction* is the first textbook to offer a straight-forward treatment of "standard" topics such as 2-dimensional manifolds, the fundamental group, and covering spaces. The author's exposition of these topics is stripped of unnecessary definitions and terminology and complemented by a wealth of examples and exercises.

Algebraic Topology: An Introduction evolved from lectures given at Yale University to graduate and undergraduate students over a period of several years. The author has incorporated the questions, criticisms and suggestions of his students in developing the text. The prerequisites for its study are minimal: some group theory, such as that normally contained in an undergraduate algebra course on the junior-senior level, and a one-semester undergraduate course in general topology.